Core Books in Advanced Mathematics

Statistics

Core Books in Advanced Mathematics

General Editor: C. PLUMPTON, Moderator in Mathematics,
University of London School Examinations Department;
formerly Reader in Engineering Mathematics,
Queen Mary College, University of London.

Titles available
Statistics
Probability
Methods of Algebra
Differentiation
Integration
Vectors
Curve Sketching
Newton's Laws and Particle Motion
Mechanics of Groups of Particles
Methods of Trigonometry
Coordinate Geometry and Complex Numbers
Proof

Core Books in Advanced Mathematics

Statistics

P. Sabine
Chief Examiner in Advanced Level Mathematics,
University of London School Examinations Department;
formerly Lecturer in Mathematics and Statistics,
Chelsea College, University of London.

C. Plumpton
Moderator in Mathematics,
University of London School Examinations Department;
formerly Reader in Engineering Mathematics,
Queen Mary College, University of London.

Macmillan Education

First published 1985

Published by
MACMILLAN EDUCATION LTD
Houndmills, Basingstoke, Hampshire RG21 2XS
and London
Companies and representatives
throughout the world

Printed in Hong Kong

ISBN 0–333–38364–8

Contents

Contents

Preface

Advanced level mathematics syllabuses are once again undergoing changes of content and approach, following the revolution in the early 1960s which led to the unfortunate dichotomy between 'modern' and 'traditional' mathematics. The current trend in syllabuses for Advanced level mathematics now being developed and published by many GCE Boards is towards an integrated approach, taking the best of the topics and approaches of the modern and traditional, in an attempt to create a realistic examination target, through syllabuses which are maximal for examining and minimal for teaching. In addition, resulting from a number of initiatives, core syllabuses are being developed for Advanced level mathematics syllabuses, consisting of techniques of pure mathematics as taught in schools and colleges at this level.

The concept of a core can be used in several ways, one of which is mentioned above, namely the idea of a core syllabus to which options such as theoretical mechanics, further pure mathematics and statistics can be added. The books in this series are core books involving a different use of the core idea. They are books on a range of topics, each of which is central to the study of Advanced level mathematics; they form small core studies of their own, of topics which together cover the main areas of any single-subject mathematics syllabus at Advanced level.

Particularly at times when economic conditions make the problems of acquiring comprehensive textbooks giving complete syllabus coverage acute, schools and colleges and individual students can collect as many of the core books as they need, one or more, to supplement books already acquired, so that the most recent syllabuses of, for example, the London, Cambridge, JMB and AEB GCE Boards, can be covered at minimum expense. Alternatively, of course, the whole set of core books gives complete syllabus coverage of single-subject Advanced level mathematics syllabuses.

The aim of each book is to develop a major topic of the single-subject syllabuses, giving essential book work and worked examples and exercises arising from the authors' vast experience of examining at this level, and also including actual past GCE questions. Thus, the core books as well as being suitable for use in either of the above ways, are ideal for supplementing comprehensive textbooks in the sense of providing more examples and exercises, so necessary for preparation and revision for examinations on the Advanced level mathematics syllabuses offered by the GCE Boards.

An attempt has been made in this book to present the work in a readable form which will encourage students to look upon statistics as an essentially experimental subject which they can apply to their studies in other subjects. A chapter on project work, with some suggested topics, has been included. There is also a chapter on the t and χ^2 tests, introducing them simply as useful tools, with appropriate conditions for their use but without mathematical derivation.

In several of the chapters, a knowledge is assumed of the probability distributions described in *Probability* in this series.

P. Sabine
C. Plumpton

Notation

$E(X)$	the mean (expectation, expected value) of the random variable X
X, Y, R, etc.	random variables
x, y, r, etc.	values of the random variables X, Y, R, etc.
$x_1, x_2, \ldots$	observations
$f_1, f_2, \ldots$	frequencies with which the observations x_1, x_2, ... occur
$\text{Var}(X)$	variance of the random variable X
$p(x)$	probability function $P(X=x)$ of the discrete random variable X
$p_1, p_2, \ldots$	probabilities of the values x_1, x_2, ... of the discrete random variable X
$f(x), g(x), \ldots$	the value of the probability density function of a continuous random variable X
$F(x), G(x), \ldots$	the value of the cumulative distribution function $P(X \leqslant x)$ of a continuous random variable X
μ	population mean
σ^2	population variance
σ	population standard deviation
$\bar{x}$	sample mean
$\hat{\sigma}^2$	unbiased estimate of a population variance from a sample size n,

$$\hat{\sigma}^2 = \frac{1}{n-1} \sum (x - \bar{x})^2$$

$B(n, p)$	binomial distribution with parameters n and p
$N(\mu, \sigma^2)$	normal distribution with mean μ and variance σ^2
ϕ	probability density function of the standardised normal variable with distribution $N(0, 1)$
Φ	cumulative distribution function of the standardised normal variable with distribution $N(0, 1)$
ρ	product–moment correlation coefficient for a population
r	product–moment correlation coefficient for a sample
r_S	Spearman's rank correlation coefficient
r_K	Kendall's rank correlation coefficient

1 Frequency distributions and the analysis of sample data

1.1 Introduction

We can describe statistics as the scientific study of numerical data. It originated in the seventeenth century, and arose from two main roots. One of these was the collection of quantitive data by government or state (hence, statistics) when interest arose in problems of census, mortality etc. for tax, insurance and other purposes. The leading figures in this work were John Gaunt (1620–74) and William Petty (1623–87). At about the same time the second root, the idea of the theory of probability, arose in connection with the great interest at that time in games of chance. Together these two aspects form the basis of modern statistics, which no longer concerns itself simply with the collection and presentation of data but deals also with the ideas of making decisions and of basing inferences on the results of the analysis of limited quantities of data.

1.2 Collection of data

The characteristic which is being recorded is called the *variate* or *variable*; it may be *discrete* (for example, when anything is counted, such as the numbers of television sets sold) or *continuous* (for example, heights and weights). When the variable being recorded is continuous, the question sometimes arises 'To how many significant figures should we record our measurements?' The measuring equipment available will, of course, partly determine the answer to this question, but, given this restriction, a rough rule is that, 'There should be at least thirty unit steps from the smallest to the largest measurement.' This means that, for example, if we are measuring the lengths of ladybirds to the nearest 1 mm and we find that we have a smallest measurement of 3 mm and a largest of 8 mm, then we have only 5 × 1 mm steps. Hence it would be better to measure, if possible, to say 0·1 mm. In this case, our extreme measurements might then be recorded as, say, 3·1 mm to 8·2 mm, which is 51 × 0·1 mm steps, and this fits the rough rule stated above.

1.3 Frequency tables

We now have our collected data, which are not generally in numerical order. Our first step is to draw up a *frequency table*. To do this we list all the possible outcomes of the variable, from smallest to largest, in one line, with the frequency of occurrence, f, of each in another line.

Example 1 The numbers of television sets sold in a given month by 30 branches of a store are shown below. Draw up a frequency table for these data.

$$
\begin{array}{cccccccccccccc}
16 & 22 & 21 & 24 & 27 & 16 & 15 & 18 & 21 & 21 & 23 & 24 & 25 & 20 & 22 \\
23 & 20 & 16 & 18 & 25 & 27 & 23 & 21 & 22 & 22 & 20 & 21 & 24 & 20 & 17
\end{array}
$$

Number of sets	15	16	17	18	19	20	21	22	23	24	25	26	27	Total
Frequency, f	1	3	1	2	0	4	5	4	3	3	2	0	2	30

(*Note*: We use this particular frequency table later to help us draw a bar diagram and also a frequency polygon.)

You will notice that, although the numbers 19 and 26 do not occur in the raw data, they should appear, with $f = 0$, in the frequency table. Otherwise, we may be misled when illustrating the data pictorially. Also, the *total frequency* must, of course, be 30.

In this example we have a frequency table for a discrete variable. When we are concerned with a continuous variable as, for example, in the measurement of the ladybirds, we will have 52 possible outcomes (from 3·1 mm to 8·2 mm) to list, for many of which f will be zero. For presentation purposes it may be advantageous to *group* the data into equal-width *class intervals*. In practice, not less than 5 and not more than 20 classes are generally used to cover the whole range of the variable. Ideally, the *boundaries* of the class intervals should not fall on an actual reading since then we would not know into which class to put that reading. This can be achieved by taking the boundaries midway between possible readings. For the ladybirds, we could take class intervals shown in the table:

	Class mark
3·05 to 3·35 mm	3·2
3·35 to 3·65 mm	3·5
3·65 to 3·95 mm	3·8
and so on.	etc.

Since each reading is to the nearest 0·1 mm, no reading can fall on a class boundary. The frequency, f, for each class is the total number of readings occurring in that class. The centre point of each class interval is called the *class mark*, and this is used to represent the class in later calculations. Thus, grouping of this kind will lead to some loss of accuracy in any calculations we may make later since we take all the readings in a class as lying at the class mark instead of their being spread throughout the class as they are in fact. With the advent of computers, grouping is becoming unnecessary for calculation purposes since a computer can handle raw sample data easily and accurately however large in number it may be.

Exercise 1.3

1 When measuring the wing lengths of a certain species of insect, it is found that the smallest collected specimen has a length of approximately 2·7 mm and the largest a length of approximately 3·4 mm. Find a suitable value of the precision with which you could measure the wing lengths of the collected insects when recording the data for future analysis.

2 A sample of size 24 from a population of snail shell lengths, in centimetres, is shown below. Draw up a frequency table for the data.

$$4·5 \quad 3·3 \quad 3·9 \quad 4·1 \quad 3·5 \quad 3·9 \quad 4·2 \quad 3·6 \quad 3·9 \quad 4·2 \quad 3·6 \quad 4·1$$
$$3·8 \quad 4·0 \quad 4·3 \quad 3·7 \quad 4·0 \quad 4·3 \quad 3·8 \quad 4·0 \quad 4·4 \quad 3·8 \quad 4·0 \quad 4·1$$

3 Given 200 measurements ranging from 1·30 mm to 2·93 mm, show how you would group them into a frequency distribution giving both the class limits and the class marks.

1.4 Pictorial illustration

The frequency table gives us some idea of the manner in which the data are arranged. Certainly we can obtain a better idea of the data as a whole than we could with the raw data. A suitable pictorial illustration will give us a more immediate appreciation of its distribution. We must be careful
 (i) to choose the right type of diagram for our particular data, and
(ii) to make sure that scales on both axes, variable and f, are marked and uniform. A diagram without a scale is useless, and, by adjusting a vertical scale, very misleading 'results' can be implied as we shall show later.
   Many types of diagrams or graphs are used in statistics; we describe those most commonly used.

Pie-chart

To construct a pie-chart we use the whole area of a circle to represent the total frequency. We then mark off sectors to represent the frequencies of the individual readings, the area of each sector being proportional to the frequency it represents. Since the area of a sector is given by $\frac{1}{2}r^2\theta$, where r is the circle radius and θ is the angle in the sector, we can construct the sectors by dividing the 360° angle at the centre of the circle into angles which are proportional to the individual frequencies.

Example 2   The method of transport to school of 1800 children in a town is shown below. Construct a pie-chart to illustrate the data.

$$\text{Car: } 300 \quad \text{Walk: } 500 \quad \text{Cycle: } 400 \quad \text{Bus: } 600$$

For car: $\dfrac{\text{angle in sector}}{360°} = \dfrac{300}{1800} \Rightarrow$ angle in sector $= 60°$.

For cycle: $\dfrac{\text{angle in sector}}{360°} = \dfrac{400}{1800} \Rightarrow$ angle in sector $= 80°$.

For walk: $\dfrac{\text{angle in sector}}{360°} = \dfrac{500}{1800} \Rightarrow$ angle in sector = 100°.

Hence for bus: angle in sector = 360° − 240° = 120°.

Check: $\dfrac{600}{1800} \times 360° = 120°$.

The pie-chart is shown in Fig. 1.1

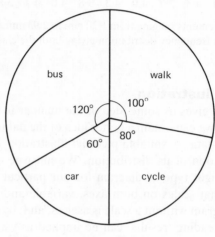

Fig. 1.1

Bar diagram

This can be used to illustrate frequencies of a discrete variable or to illustrate frequencies of attributes such as 'car', 'cycle', 'walk', and 'bus' of our previous example. A bar is drawn for each value of the variable, or for each attribute, such that the length of the bar is proportional to the frequency of that value of the variable or of that attribute. The bars are usually thickened to make them more easily seen; the actual thickness is of no significance but all the bars should be given the same thickness. However, the bars should be kept distinct, not joined up, to indicate that the variable is discrete.

Example 3 Draw a bar diagram to illustrate the data (on the number of television sets sold) for which we constructed a frequency table in §1.3.

Here (Fig. 1.2) the axes are carefully marked but it would be foolish to start at zero the 'Number of television sets' axis and then to have a long gap before the first reading at 15. You will notice that we have put a sign, $\rightsquigarrow$, just to the right of the origin to indicate that this axis is 'broken', and then we start at the entry 15. However, this must *never* be done on the frequency axis for then the data would be misrepresented. For example, compare the two simple

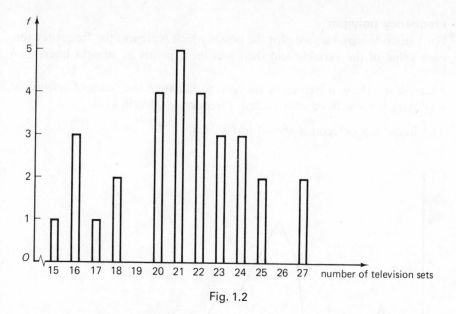

Fig. 1.2

diagrams of Fig. 1.3; (a) is correct but (b) is incorrect and gives us a totally misleading idea of the data.

Given:

Number of items	1	2
f	20	30

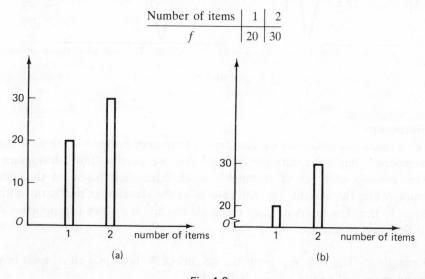

Fig. 1.3

This precept applies equally to all the following pictorial representations which we describe.

Frequency polygon

For a discrete variable, we plot the points which represent the frequency for each value of the variable and then join these points by straight lines.

Example 4 Draw a frequency polygon to illustrate the 'sales of television sets' data for which we constructed a frequency table in §1.3.

The frequency polygon is shown in Fig. 1.4.

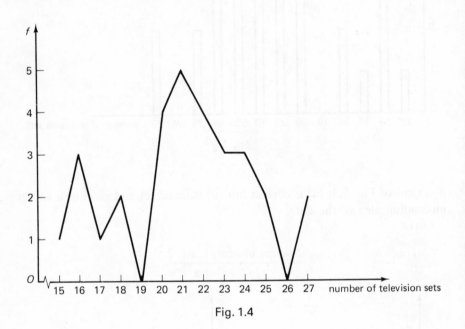

Fig. 1.4

Histogram

For a continuous variable we can draw a frequency polygon if the data are ungrouped, but if the data are grouped then we usually draw a histogram. This consists of a set of rectangles which have their bases on the axis representing the variable, x, with centres at the class marks and with width equal to the class interval. Each rectangle has an area which is proportional to the class frequency.

Example 5 The data represent the heights of 30 boys in a class, each boy being measured to the nearest half-inch.

Height (inches)	$48\frac{1}{4}$–$52\frac{1}{4}$	$52\frac{1}{4}$–$56\frac{1}{4}$	$56\frac{1}{4}$–$60\frac{1}{4}$	$60\frac{1}{4}$–$64\frac{1}{4}$	$64\frac{1}{4}$–$68\frac{1}{4}$	$68\frac{1}{4}$–$72\frac{1}{4}$
f	7	14	6	2	0	1

Draw a histogram to illustrate the data.

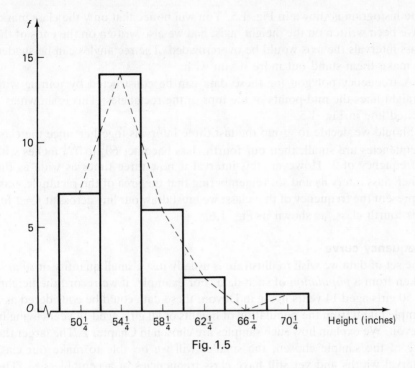

Fig. 1.5

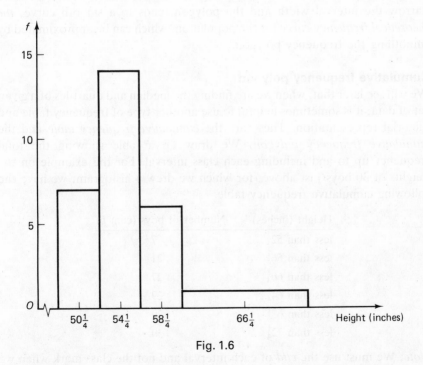

Fig. 1.6

The histogram is shown in Fig. 1.5. You will notice that only the class marks have been written on the 'height' axis; had we also written on the ends of the class intervals the axis would be overcrowded. The rectangles can be shaded to make them stand out more if you wish.

A frequency polygon for these data can be constructed by joining with straight lines the mid-points of the tops of the rectangles. This is shown as a dotted line in Fig. 1.5.

Should we decide to group the last three intervals together since the class frequencies are small, then our fourth class becomes $60\frac{1}{4}$ to $72\frac{1}{4}$ inches with a frequency of 3. However, this interval is now three times as wide as the other class intervals and so, remembering that the area of the rectangle must represent the frequency of that class, we must draw our line across at $f = 1$ for this fourth class, as shown in Fig. 1.6.

Frequency curve

The set of data we wish to illustrate is usually just a small quantity or *sample* taken from a *population* of such data. For example, if we record the heights of 50 girls aged 14 years living in Devon, these data could be considered as a sample taken from the population of heights of all girls aged 14 years living in Devon. We explain how such samples are chosen in Chapter 2. The larger the size of the sample chosen, the smaller will we be able to make our class interval widths and yet still have class frequencies of acceptable size. The straight line joins on the frequency polygon become shorter and shorter as we narrow the interval width and the polygon tends to a smooth curve, *the theoretical frequency curve* for the population, which can be approximated by smoothing the frequency polygon.

Cumulative frequency polygon

We will see later that, when we are finding the median and quartiles of a given set of data, it is sometimes helpful to use another type of frequency table and pictorial representation. These are the *cumulative frequency table* and the *cumulative frequency polygon*. We draw up a table showing the total frequency up to and including each class interval. For the example on the heights of 30 boys just above, for which we drew a histogram, we have the following cumulative frequency table:

Height (inches)	Number of boys (cum f)
less than $52\frac{1}{4}$	7
less than $56\frac{1}{4}$	21
less than $60\frac{1}{4}$	27
less than $64\frac{1}{4}$	29
less than $68\frac{1}{4}$	29
less than $72\frac{1}{4}$	30

Note: We must use the *end* of each interval and not the class mark when we are forming this frequency table, for we are recording the number of boys up to and including the whole of a particular class. Also, the final cum f must be equal to the total frequency of the sample (30 in our example).

The points $(52\frac{1}{4}, 7)$, $(56\frac{1}{4}, 21)$, $(60\frac{1}{4}, 27)$, $(64\frac{1}{4}, 29)$, $(68\frac{1}{4}, 29)$, $(72\frac{1}{4}, 30)$ are plotted and joined by straight lines to form a *cumulative frequency polygon* or *ogive*. This is shown in Fig. 1.7.

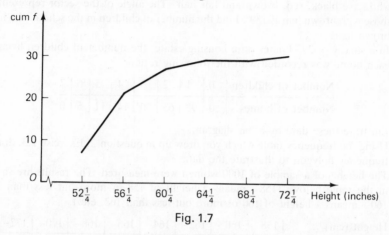

Fig. 1.7

We may sometimes wish to draw a cumulative frequency polygon for the cum f's of all values *greater than or equal to* the lower class-boundary of each class interval. The graph would then take the shape of Fig. 1.8.

Height (inches)	cum f
$\geqslant 48\frac{1}{4}$	30
$\geqslant 52\frac{1}{4}$	23
$\geqslant 56\frac{1}{4}$	9
$\geqslant 60\frac{1}{4}$	3
$\geqslant 64\frac{1}{4}$	1
$\geqslant 68\frac{1}{4}$	1

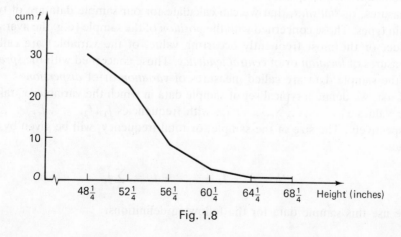

Fig. 1.8

Frequency distributions and the analysis of sample data 9

As in the case for a frequency polygon, a theoretical *cumulative frequency curve* for the population from which the sample comes can be approximated by smoothing a cumulative frequency polygon.

Exercise 1.4

1 A pie-chart is drawn showing the numbers of children in a school of 600 pupils who have black, red, brown and fair hair. The angle of the sector representing those with brown hair is 150°. Find the number of children in the school who have brown hair.

2 In a survey of 295 homes on a housing estate, the number of children living in each home was recorded with the following results:

Number of children	0	1	2	3	4	5	6	7
Number of homes	50	92	63	60	14	11	5	0

Illustrate these data by a bar diagram.

3 Using the frequency table which you drew up in question 2, Exercise 1.3, draw a frequency polygon to illustrate the data.

4 The heights of a sample of 1000 women were measured. The results are shown in the table below. (158– means a height of 158 or more but less than 160, 160– means a height of 160 or more but less than 162, etc.)

Height (cm)	158–	160–	162–	164–	166–	168–	170–	172–
Number of women	12	35	82	105	148	141	125	136

Height	174–	176–	178–	180–	182 and above
Number of women	98	66	40	12	0

Draw a histogram to illustrate these data.

Construct a cumulative frequency table and draw a cumulative frequency polygon.

Sketch by eye an estimate of the theoretical cumulative frequency curve for the population from which this sample comes.

1.5 Statistics of sample data

We wish to derive more information from the sample data than can be obtained from a frequency table and a suitable pictorial representation. The measures, or *statistics*, that we can calculate for our sample data are of two main types. Those concerned with the *position* of the sample (e.g. the average value, or the most frequently occurring value, of the variable) are called measures of *location* or of *central tendency*. Those concerned with the *spread* of the sample data are called measures of *variation* or of *dispersion*.

First, we define a typical set of sample data in which the variable, x, takes the values $x_1, x_2, \ldots, x_r, \ldots, x_n$, with frequencies $f_1, f_2, \ldots, f_r, \ldots, f_n$, respectively. The size of the sample, or total frequency, will be given by N, where

$$N = f_1 + f_2 + \ldots + f_n = \sum_{r=1}^{n} f_r.$$

We use this sample data for the following definitions.

Measures of location

The (arithmetic) mean, $\bar{x}$ (called 'x bar'), of the set of readings (or observations) is defined by

$$\bar{x} = \frac{x_1 f_1 + x_2 f_2 + \ldots + x_r f_r + \ldots + x_n f_n}{f_1 + f_2 + \ldots + f_r + \ldots + f_n}$$

$$= \sum_{r=1}^{n} x_r f_r \Big/ \sum_{r=1}^{n} f_r$$

$$= \sum_{r=1}^{n} x_r f_r \Big/ N.$$

Example 6 Find the mean of the following data:

x	1	3	5	7	9
f	2	0	4	3	1

Total frequency $= 10$, $\bar{x} = \dfrac{2 + 0 + 20 + 21 + 9}{10}$

$$= 5 \cdot 2.$$

Example 7 Find, to three decimal places, the mean of the following grouped data.

x	10–15	15–20	20–25	25–30
f	4	8	12	5

First we must write down the class marks. These are

$$12\tfrac{1}{2} \quad 17\tfrac{1}{2} \quad 22\tfrac{1}{2} \quad 27\tfrac{1}{2}.$$

$$\bar{x} = \frac{4 \times 2\tfrac{1}{2} + 8 \times 17\tfrac{1}{2} + 12 \times 22\tfrac{1}{2} + 5 \times 27\tfrac{1}{2}}{29} = \frac{597 \cdot 5}{29}$$

$$= 20 \cdot 603 \approx 20 \cdot 6.$$

Note: Later we will use these data to find the variance and the standard deviation.

The weighted (arithmetic) mean. Sometimes certain readings in the sample data have more importance or significance than others, and we wish the measure of the mean value to take account of this. We attach *weights* w_1, w_2, $\ldots$, w_n to the readings x_1, x_2, $\ldots$, x_n and define the *weighted mean* as

$$\frac{w_1 x_1 + w_2 x_2 + \ldots + w_n x_n}{w_1 + w_2 + \ldots + w_n}.$$

Example 8 In an honours degree course, the average mark of a student at the June examination is weighted according to the year of study. The weighting is 1 for the first year, 2 for the second year, and 3 for the third year.

A student has marks of 70%, 52%, 55% respectively in the three years of study. Find her weighted mean percentage mark.

$$\text{Weighted mean percentage} = \frac{70 \times 1 + 52 \times 2 + 55 \times 3}{1 + 2 + 3}\% = \frac{339}{6}\%$$
$$= 56.5\%.$$

The mode of a set of readings is the value of the variable which has the largest frequency.

In Example 6 just above we have a mode of 5. In example 7 just above we have a *modal class*, the class 20 to 25.

The mode may not be unique (there may be two or more values of the variable with the largest frequency value). A frequency distribution is said to be *unimodal* if it has only one mode.

The median, M, of a set of *N* readings is the value taken by the middle reading when the set has been placed in order of size. When *N* is an odd number, this will be the value of the $\frac{N + 1}{2}$ th reading. When *N* is an even number, the median is taken as half the sum of the *N*/2 th and (*N* + 2)/2 th readings.

Example 9 Find the median of the set of readings

(a) 2 5 4 7 1 8 5, (b) 2 5 4 7 1 8.

(a) In ascending order we have 1 2 4 5 5 7 8. That is, 7 readings. The median is the $\frac{7 + 1}{2}$ = fourth reading

$\Rightarrow$ Median = 5.

(b) In ascending order we have 1 2 4 5 7 8. That is, 6 readings. Here we have 2 middle readings (the third and fourth)

$\Rightarrow$ Median = $\frac{4 + 5}{2}$ = 4·5.

You will notice that the median is not concerned with the actual sizes of the readings in the sample, apart from the middle reading(s). It is therefore little affected by extreme values in the set of data. For example, the two sets of readings:

1 20 21 23   and   19 20 21 24

have the same median values (but very different means). For this reason, the median is sometimes preferred to the mean as a representative 'average' value of a set of data when the data include one or two extreme values. For example, when considering the 'average' salary of the staff of a particular office in which there are thirty staff with salaries ranging from £5000 to £12000 per annum and one manager earning £20000, the mean salary would be considerably affected by the one salary of £20000, whereas the median would be much less affected by this extreme value. Thus, in this example, the median would be more representative of an 'average salary' than would be the mean. The median is generally preferred to the mean as a representative 'average' when the distribution of the data is far from symmetrical about the mean, that is, when the distribution *is skewed*.

We can also find the median from the cumulative frequency graph by finding the value for x when cum $f = (N + 1)/2$.

Example 10 Draw the cumulative frequency (cum f) polygon for the following sample data and hence find the median value.

x	1	2	3	4	5	6
f	2	6	7	4	2	1

The cumulative frequency table is

x	$\leqslant 1$	$\leqslant 2$	$\leqslant 3$	$\leqslant 4$	$\leqslant 5$	$\leqslant 6$
cum f	2	8	15	19	21	22

The polygon is shown in Fig. 1.9.

$$N = 22 \Rightarrow \frac{N + 1}{2} = 11{\cdot}5.$$

We draw a line across to the polygon at cum $f = 11{\cdot}5$. The corresponding x reading, $2{\cdot}5$, is the median.

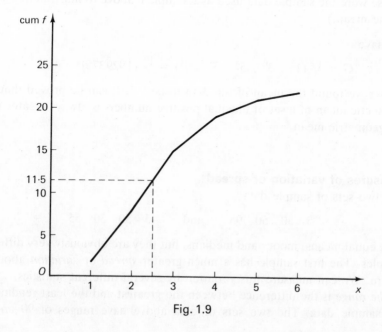

Fig. 1.9

We often wish to find other *quantiles* of our data. A quantile is the value of the variable such that a given fraction of the total amount of data lies to its left. The median is the value of x which has 50% of the data on either side of it. The *quartiles* divide the data set into four equal parts; thus, the first quartile has 25% of the data to its left, the second quartile is the median and the third quartile has 75% of the data to its left.

Other similar divisions used are *deciles*, division into tenths, and *percentiles*, division into hundredths. All the quantiles can be found, either from the data or from the cum f graph by the same methods by which we found the median. In Fig. 1.9 the first and third quartiles can be found by finding the x values corresponding to cum f equal to $\dfrac{N + 1}{4} = 5 \cdot 75$ and $\dfrac{3(N + 1)}{4} = 17 \cdot 25$, respectively. The difference between the third and first quartiles is called the *interquartile range*.

The geometric mean, G, is defined as

$$G = \sqrt[N]{[\ (x_1)^{f_1} \cdot (x_2)^{f_2} \ \ldots \ (x_n)^{f_n}]},$$

the positive real root being taken.

Example 11   Find, to two decimal places, the geometric mean for the sample data

x	1	3	5	7	9
f	2	0	4	3	1

(These were the sample data used as Example 6 above to illustrate the work on the mean.)

We have

$$G = \sqrt[10]{(1^2 \cdot 3^0 \cdot 5^4 \cdot 7^3 \cdot 9^1)} = \sqrt[10]{(1\,929\,375)} \approx 4 \cdot 25.$$

Earlier we found the mean of the data to be $5 \cdot 2$. It can be proved that the arithmetic mean of a set of unequal positive numbers is always greater than the geometric mean.

Measures of variation or spread

The two sets of sample data

$$5 \quad 50 \quad 50 \quad 95 \quad \text{and} \quad 45 \quad 50 \quad 50 \quad 55$$

have equal means, modes and medians, but they are obviously very different samples. The first sample has a much greater *spread* or *variation* about its centre. We can measure this variation by several different statistics.

The range is the difference between the greatest and the least readings in the sample data. The two sets of data above have ranges of 90 and 10 respectively. In practice, the range is often given by quoting the smallest and largest readings, i.e. (5, 95) and (45, 55), for these tell the reader of a scientific paper more about the relative spread of the data than do the ranges of 90 and 10.

The variance and the standard deviation. The difference between a typical reading, x_r, and the mean, $\bar{x}$, $(x_r - \bar{x})$, is called the *deviation* of x_r from the mean. For a given sample, some readings will have a positive deviation from the mean and others a negative deviation. From the definition of $\bar{x}$, we know that

$$\sum_{r=1}^{n} f_r (x_r - \bar{x}) = 0,$$

so we cannot use the summation on the left-hand side of this equation to define a measure of the spread of the data. A measure, the *mean deviation*, which utilises the sum of the *positive* values of all the deviations from the mean and which is defined by

$$\text{MD} = \frac{1}{N} \sum_{r=1}^{n} f_r |x_r - \bar{x}|,$$

is now little used, because it is difficult to use this summation in any further theoretical work. The principal measures that we use are the *variance* and the *standard deviation*. These use the sum of the *squares* of the deviations from the mean. We define the *variance* of the set of data as s^2, where

$$s^2 = \frac{1}{N} \sum_{r=1}^{n} f_r (x_r - \bar{x})^2,$$

and the *standard deviation, s,* by

$$\text{SD} \equiv s = \sqrt{(\text{variance})},$$

the positive square root being taken. The standard deviation will have the same units as the variable, while the variance will have the square of those units.

In numerical calculations for finding variance or SD, $(x_r - \bar{x})$ is seldom an integer; the working can sometimes be shortened by using an equivalent formula:

$$s^2 = \frac{1}{N} \sum_{r=1}^{n} f_r x_r^2 - \bar{x}^2.$$

Proof By definition,

$$s^2 = \frac{1}{N} \sum_{r=1}^{n} f_r(x_r - \bar{x})^2 = \frac{1}{N} \sum_{r=1}^{n} f_r x_r^2 - \frac{2\bar{x}}{N} \sum_{r=1}^{n} f_r x_r + \frac{\bar{x}^2}{N} \sum_{r=1}^{n} f_r$$

$$= \frac{1}{N} \sum_{r=1}^{n} f_r x_r^2 - 2\bar{x}^2 + \frac{\bar{x}^2 N}{N}$$

$$= \frac{1}{N} \sum_{r=1}^{n} f_r x_r^2 - \bar{x}^2.$$

The use of this latter formula is not recommended when the readings x_1, x_2, ..., x_n are large and hence $\bar{x}$ is large, for then results may be inaccurate due to rounding-off errors.

Example 12 Find the mean and the variance of the following data. Find also, to two decimal places, the standard deviation.

$$x: \quad 5 \quad 3 \quad 7 \quad 2 \quad 8 \quad 4 \quad 1$$

$$N = 7, \quad f_r = 1 \text{ for all } r$$

$$\Rightarrow \bar{x} = \frac{5 + 3 + 7 + 2 + 8 + 4 + 1}{7} = \frac{30}{7}.$$

$$s^2 = \frac{1}{7} (5^2 + 3^2 + 7^2 + 2^2 + 8^2 + 4^2 + 1^2) - \left(\frac{30}{7}\right)^2$$

$$= \frac{1}{7} (168) - \frac{900}{49} = \frac{276}{49}$$

$$\Rightarrow s = 2 \cdot 37.$$

Example 13 Find, to two decimal places, the variance and the standard deviation for the grouped data of Example 7 above for which we found the mean.

We have

x	f	Class mark (c)	fc^2
10–15	4	$12\frac{1}{2}$	625
15–20	8	$17\frac{1}{2}$	2 450
20–25	12	$22\frac{1}{2}$	6 075
25–30	5	$27\frac{1}{2}$	3 781·25
	29		12 931·25

Earlier we found $\bar{x} = \dfrac{1195}{58}$.

$$s^2 = \frac{12\,931 \cdot 25}{29} - \left(\frac{1195}{58}\right)^2 = \frac{72\,000}{3364} = 21 \cdot 403\,09$$

$$\Rightarrow \text{variance} = 21 \cdot 40 \text{ (to two decimal places)}$$

$$\Rightarrow \text{SD} = 4 \cdot 63 \text{ (to two decimal places)}.$$

Exercise 1.5

1 The distribution by age of the total population of 1234 people living on an island is shown below. Draw a histogram to illustrate these data and estimate the mean age of the population.

Age (years)	0–4	5–14	14–19	20–34	35–49	50–64	65–74	75–84	85+
Number	120	245	96	256	238	189	65	25	0

2 Find, to two decimal places, the geometric mean for the sample data

x	1	2	3	4	5	6
f	2	4	0	5	2	1

3 In a sixth form the numbers of days in the year that the 12 members of the class were away from school because of illness were as follows:

$$12 \quad 4 \quad 2 \quad 0 \quad 45 \quad 7 \quad 5 \quad 10 \quad 14 \quad 1 \quad 0 \quad 8.$$

Find the mean, the mode and the median for these data.

State, giving your reasons, which of these values you consider gives you the best measure of the average absence in the year due to illness of the members of the class.

4 The following data are the concentrations of a trace element (milligrams per litre) in 7 samples of water taken from a spring:

$$244 \cdot 8 \quad 241 \cdot 3 \quad 240 \cdot 7 \quad 240 \cdot 6 \quad 238 \cdot 2 \quad 237 \cdot 9 \quad 236 \cdot 5.$$

For these data (a) state the range, (b) find the mean deviation, (c) find the variance and the standard deviation.

5 For the data of question 4 Exercise 1.4, find, to 2 decimal places, the mean height. State the modal class.

6 From the cumulative frequency curve which you drew for question 4 in Exercise 1.4, estimate the median height and the upper and lower quartiles for the population of heights.

1.6 Combined mean and variance for two or more samples

We may have two or more samples of data, not necessarily of equal size, arising from the same experiment, so that we have two samples from the same population of data. If we have already calculated the mean and the variance of the separate samples, we may wish to find the mean and the variance of the combined sample of readings without recalculating the $\bar{x}$ and s^2 for the combined set. Given the sets of sample data

sample 1: $x_1, x_2, \ldots, x_{n_1}$; size n_1; mean $\bar{x}$; SD s_x;
sample 2: $y_1, y_2, \ldots, y_{n_2}$; size n_2; mean $\bar{y}$; SD s_y;
sample 3: $z_1, z_2, \ldots, z_{n_3}$; size n_3; mean $\bar{z}$; SD s_z; etc.,

then the mean of the combined sample is given, by definition, by

$$\bar{X} = \frac{x_1 + x_2 + \ldots + x_{n_1} + y_1 + y_2 + \ldots + y_{n_2} + z_1 + z_2 + \ldots + z_{n_3} + \ldots}{n_1 + n_2 + n_3 + \ldots}.$$

Since
$$x_1 + x_2 + \ldots + x_{n_1} = n_1\bar{x},$$
$$y_1 + y_2 + \ldots + y_{n_2} = n_2\bar{y},$$
$$z_1 + z_2 + \ldots + z_{n_3} = n_3\bar{z}, \text{ etc.},$$
$$\bar{X} = \frac{n_1\bar{x} + n_2\bar{y} + n_3\bar{z} + \ldots}{n_1 + n_2 + n_3 + \ldots}.$$

By definition, the variance of the combined sample is given by

$$s^2 = \frac{\sum_{r=1}^{n_1}(x_r - \bar{X})^2 + \sum_{r=1}^{n_2}(y_r - \bar{X})^2 + \sum_{r=1}^{n_3}(z_r - \bar{X})^2 + \ldots}{n_1 + n_2 + n_3 + \ldots}.$$

In general, we cannot shorten the working for this to any great extent. We could write

$$\sum_{r=1}^{n_1}(x_r - \bar{X})^2 = \sum_{r=1}^{n_1}[(x_r - \bar{x}) + (\bar{x} - \bar{X})]^2$$

$$= \sum_{r=1}^{n_1}(x_r - \bar{x})^2 + 2(\bar{x} - \bar{X})\sum_{r=1}^{n_1}(x_r - \bar{x}) + (\bar{x} - \bar{X})^2\sum_{r=1}^{n_1}1$$

$$= \sum_{r=1}^{n_1}(x_r - \bar{x})^2 + n_1(\bar{x} - \bar{X})^2,$$

since $\sum_{r=1}^{n_1}1 = n_1$, and $\sum_{r=1}^{n_1}(x_r - \bar{x}) = 0$,

$$\Rightarrow \sum_{r=1}^{n_1}(x_r - \bar{X})^2 = n_1[s_x^2 + (\bar{x} - \bar{X})^2]$$

$$\Rightarrow s^2 = \frac{n_1 s_x^2 + n_2 s_y^2 + n_3 s_z^2 + \ldots + n_1(\bar{x} - \bar{X})^2 + n_2(\bar{y} - \bar{X})^2 + n_3(\bar{z} - \bar{X})^2 \ldots}{n_1 + n_2 + n_3 + \ldots},$$

which avoids the calculation of any further summations.

Example 14 The following data give the lengths (mm) of two samples of earwigs taken from under a large rock in a garden.

Sample 1: 8·5 9 9 10 8
Sample 2: 11 9 9·5 10·5

Find (a) the mean earwig length and its standard deviation for each sample,
(b) the mean and the variance of the combined sample of earwig lengths, giving your answers to three decimal places.

(a) For sample 1, $\bar{x}_1 = \dfrac{8\cdot5 + 9 + 9 + 10 + 8}{5}$ mm $= 8\cdot9$ mm.

$$s_1^2 = \dfrac{(0\cdot4)^2 + (0\cdot1)^2 + (0\cdot1)^2 + (1\cdot1)^2 + (0\cdot9)^2}{5} \text{ (mm)}^2$$

$$= \dfrac{2\cdot2}{5} \text{ (mm)}^2 = 0\cdot44 \text{ (mm)}^2.$$

$$s_1 = 0\cdot663 \text{ mm.}$$

For sample 2, $\bar{x}_2 = \dfrac{11 + 9 + 9\cdot5 + 10\cdot5}{4}$ mm $= 10$ mm.

$$s_2^2 = \dfrac{1^2 + 1^2 + (0\cdot5)^2 + (0\cdot5)^2}{4} \text{ (mm)}^2$$

$$= 0\cdot625 \text{ (mm)}^2.$$

$$s_2 = 0\cdot791 \text{ mm.}$$

(b) Combined mean, $\bar{X} = \dfrac{5 \times 8\cdot9 + 4 \times 10}{9}$ mm $= 9\cdot389$ mm.

Combined variance,

$$s^2 = \dfrac{5 \times 0\cdot44 + 4 \times 0\cdot625 + 5\,(9\cdot389 - 8\cdot9)^2 + 4\,(10 - 9\cdot389)^2}{9} \text{ (mm)}^2$$

$$= \dfrac{7\cdot3889}{9} \text{ (mm)}^2$$

$$= 0\cdot821 \text{ (mm)}^2.$$

Or combined variance,

$$s^2 = \dfrac{\begin{array}{c}(8\cdot5) - 9\cdot389)^2 + 3(9 - 9\cdot389)^2 + (10 - 9\cdot389)^2 + (8 - 9\cdot389)^2 + \\ (11 - 9\cdot389)^2 + (9\cdot5 - 9\cdot389)^2 + (10\cdot5 - 9\cdot389)^2\end{array}}{9} \text{ (mm)}^2$$

which gives the same result.

Exercise 1.6

1 Ten tomato plants were grown from seed and their heights after one month under identical conditions were recorded, giving the following results:

Height (cm): 4 5 8 3 2 6 10 8 5 7.

Find the mean height and its standard deviation.
Another eight plants, grown under the same conditions for the same length of time, were also measured. Their heights were:

Height (cm): 6 8 4 9 3 6 7 5.

Find the mean height and its standard deviation for these eight plants.
Find also the mean height and the variance of the combined set of eighteen plants, giving your answers to two decimal places.

1.7 Index numbers

An index number is a measure which is used to show the change in a variable with respect to some particular characteristic. Hence, an index number will be given relative to some starting point or *base*. For example, we may require an index number which will show us the change in the cost of living over a certain period of years, or we may need one to compare the total production of tin in a given year by two Cornish tin mines. Index numbers can be used for forecasting future results in industry, education, government, etc. Many of the indices which we see and hear quoted show changes with respect to time, but the formulae quoted below hold equally well for changes with respect to location, or type of work or other characteristic.

To find an index number representing the change in the cost of living, we need a weighted index in which the weighting is determined by the average quantity of each item purchased. Suppose that N basic items are being considered, and that the prices for those items in the *base year* and in the year under consideration are known, as are the average quantities purchased in the year under consideration. The *Paasche* index is given by

$$\frac{\Sigma(\text{price in the year under consideration}) (\text{average quantity purchased in the year under consideration})}{\Sigma(\text{price in the base year}) (\text{average quantity purchased in the year under consideration})},$$

the summations being taken over all the items.

We define two more indices although there are many others which can be used. The *Laspeyres* index has the same form as the Paasche index except that the 'average quantities purchased' which are used are those for the base year only, so that the definition becomes

$$\frac{\Sigma(\text{price in the year under consideration}) (\text{average quantity purchased in the base year})}{\Sigma(\text{price in the base year}) (\text{average quantity purchased in the base year})},$$

the summations being taken over all the items.

Fisher's index is the geometric mean of the Paasche and the Laspeyres indices:

$$\text{Fisher's index} = \sqrt{[(\text{Paasche index})(\text{Laspeyres index})]}.$$

None of these indices is ideal, although all are suitable for use in practice. Theoretically, Fisher's index is the best of the three.

Example 15 The table shows the average price, in pence per kilo, of six staple food items for the years 1980 and 1984. It also shows the average weight in kilos consumed per person in those years.

Item	1980 Price	1980 Weight	1984 Price	1984 Weight
Bread	35	52	55	46
Butter	100	20	124	15
Margarine	100	18	122	24
Potatoes	20	54	30	56
Sugar	50	20	60	18
Tea	100	14	140	10

Using 1980 as the base year, find an index number for the year 1984 for these staple foods.

$$\text{Paasche index} = \frac{55 \times 46 + 124 \times 15 + 122 \times 24 + 30 \times 56 + 60 \times 18 + 140 \times 10}{35 \times 46 + 100 \times 15 + 100 \times 24 + 20 \times 56 + 50 \times 18 + 100 \times 10}$$

$$= \frac{11\,478}{8530} = 1 \cdot 346.$$

$$\text{Laspeyres index} = \frac{55 \times 52 + 124 \times 20 + 122 \times 18 + 30 \times 54 + 60 \times 20 + 140 \times 14}{35 \times 52 + 100 \times 20 + 100 \times 18 + 20 \times 54 + 50 \times 20 + 100 \times 14}$$

$$= \frac{12\,316}{9100} = 1 \cdot 353.$$

Fisher's index $= \sqrt{(1 \cdot 346 \times 1 \cdot 353)} = 1 \cdot 349.$

Each of these three indices indicates to us that, if 1980 is taken as 100%, then the 1984 price index for these basic foods is approximately 135%.

Exercise 1.7

1 A factory uses four raw materials A, B, C and D to produce salad cream. The masses of the materials used in the production are in the ratios $1:2:4:3$ respectively. The prices of the materials in pounds sterling per tonne in the years 1982 and 1984 are given below:

	A	B	C	D
1982	6	4	3	1
1984	8	5	5	2

Calculate an index number for the total cost of the materials used to manufacture the salad cream in 1984 using 1982 as the base year.

Assuming that the selling price is directly proportional to the cost of the materials used, find the selling price of the salad cream in 1984, given that the 1982 selling price was 85p.

Miscellaneous Exercise 1

1 A trap for catching flies is set in a kitchen and the flies caught are removed and counted each evening on 20 consecutive days. The number of flies caught each day is shown below:

```
32   5  20  15  22  14  11  21  19  12
10   6  22  16  28  11  21  22  20  10
```

Find the mean number of flies caught per day with its standard deviation.
Find also the median and the mode for the data.

2 An educationalist gave a test to 400 children. The marks gained in the test are shown in the table:

Mark (x)	$60 \leq x < 80$	$80 \leq x < 100$	$100 \leq x < 120$	$120 \leq x < 140$	$140 \leq x < 160$
Number of children	61	83	107	79	70

Write down the class mark for each of the intervals.
Estimate the mean mark and the variance for the sample.

3 A geography paper was sat by 50 candidates who scored, out of a total of 100, the marks shown below. All marks awarded were integers. Draw a histogram to illustrate the distribution of the marks.
Estimate the mean mark, explaining the limitations of your calculation.
Draw a cumulative frequency polygon. Estimate the median mark and the inter-quartile range.
State, giving your reasons, whether you consider the mean or the median to be the better measure of the average performance of these candidates.

Mark range	<30	30–39	40–49	50–59	60–69	70–79	80 and over
Frequency	1	6	8	10	18	5	2

4 The table shows the times, t seconds, measured to the nearest 0·1 second, taken for sugar cubes to dissolve in water at a given temperature. Draw a histogram to represent the data.
Estimate the mean time, explaining the limitations of your calculations.
Draw a cumulative frequency diagram for the data. Estimate the median and the other two quartiles.

Time, t (seconds)	Number of cubes
$t < 20$	40
$20 \leq t < 21$	145
$21 \leq t < 22$	286
$22 \leq t < 22 \cdot 5$	200
$22 \cdot 5 \leq t < 23$	252
$23 \leq t < 23 \cdot 5$	310
$23 \cdot 5 \leq t < 24$	201
$24 \leq t < 24 \cdot 5$	90
$24 \cdot 5 \leq t < 25$	50
$t \geq 25$	6

5 Three samples of a type of spring being produced by a factory were taken, the samples consisting of 50, 30 and 20 springs respectively. A weight was attached to each spring and the percentage extension measured. The results for each of the three samples are summarised below:

Sample number	Size of sample	Mean	SD
1	50	12·7	1·15
2	30	13·2	0·95
3	20	14·5	1·25

Find the mean percentage extension and its standard deviation for the combined sample of 100 springs.

6 Fifteen professional and ten amateur skaters took part in a skating competition, each being awarded a mark which could range from 0·0 to 6·0 (each mark was given to one decimal place). A summary of the results obtained is given below:

	Mean score	Standard deviation
15 professionals	4·3	0·8
10 amateurs	3·2	1·2

Calculate the mean and the standard deviation of the combined set of twenty-five scores.

One of the skaters fell, injuring a knee, during the competition. As a result, he scored only 0·7 marks. It was decided to exclude his result. Calculate the mean and the standard deviation of the other twenty-four scores.

7 For the following grouped data, calculate the mean, the standard deviation and the mean deviation:

Class mark	95	105	115	125	135	145	155	165	175
Frequency	3	8	25	40	38	15	10	7	4

8 The table shows the average wholesale prices and the total production in the UK for three basic foodstuffs for the years 1981 and 1984. Using 1981 as the base year, calculate (a) Laspeyres, (b) Paasche, (c) Fisher price indices for the year 1984. In each case give your answer as a percentage to two decimal places.

Foodstuff	Price (pence per lb) 1981	1984	Production (million lb) 1981	1984
1	3·95	4·15	9680	10436
2	61·5	59·7	117·7	115·5
3	34·8	38·9	77·93	82·79

9 Given that the weighted mean cost in 1980 of four raw materials A, B, C, D, used in the production of an article, is 1·00, it is found that the weighted mean cost in 1981 is 1·09, and in 1982 it is 1·20. In 1981, the cost of A had risen by 9% and the cost of B by 8% from their 1980 values. In 1982 their costs had risen by 18% and 20% respectively from their 1980 values. The cost of material C is known to rise by $x\%$ each year and the cost of D is known to rise by $y\%$ each year. The four materials have weights in the ratios $2:4:1:3$ respectively. Assuming that the costs of the four raw materials were equal in 1980, find values for x and y.

10 A set of N readings $x_1, x_2, \ldots, x_N$ has mean $\bar{x}$ and standard deviation s_x. Another set of M readings $y_1, y_2, \ldots, y_M$ has mean $\bar{y}$ and standard deviation s_y. Given that the mean of the combined set of $(N + M)$ readings has mean m with standard deviation s, show that

(a) $(N + M)m = N\bar{x} + M\bar{y}$,

(b) $(N + M)(s^2 + m^2) = N(s_x^2 + \bar{x}^2) + M(s_y^2 + \bar{y}^2)$.

The mean of 15 measurements is 1·34 with standard deviation 0·90. A further measurement of 1·50 is then included. Find the mean and the standard deviation of the set of 16 measurements.

2 Random samples and sampling distributions

2.1 Sample and population

In Chapter 1 we dealt with the frequency distribution of a sample; such a set of sample data is a subset of the whole population of such data. For example, the managers of a bus company may wish to discover the average journey time of passengers on the company buses. They cannot afford the expense and time necessary to ask every passenger, but they can ask a *sample* of passengers to state their journey times and, from this sample, the company will try to make inferences about the whole *population* of journey times.

A population may be finite or infinite. It is very important that a sample is chosen carefully if valid inferences about the population are to be made from it, for, ideally, we wish the sample to be 'typical' of the population. We do not want the sample to be chosen in a subjective way, with any personal or other bias.

Such bias can be avoided by choosing a *random sample* from the population; that is, we choose our sample so that every member of the population has an equal chance of being chosen as a member of the sample. Where it is possible to give every member of the population a number, or a pair of numbers, a *table of random sampling numbers* can be used to help in selecting the sample. These tables consist of lists of digits from 0 to 9 inclusive which have been generated randomly, that is, so that each digit has a probability of 1/10 of occurring. Nowadays computers can be used to generate the listed numbers.

Let us suppose that we wish to choose a random sample of 100 cabbages from a large field of cabbages. We could consider each point in the field as having two coordinates referred to axes along adjacent sides of the field. Then we could choose any row or column we wish from a table of random numbers; for example, if we decide to start by using the sixth row of our table, we might find that row is

    1174   2693   8144   3393   0872   3279   7331   1822   6470   6850.

Provided that the lengths of the sides of the field are less than 100 m, we could take sample cabbages at the coordinates (m) (11,74), (26,93), (81,44) and so on until we get to the end of that row when we move to other rows in the table and proceed in the same way until the 100 sample cabbages are chosen. We reject any numbers which fall outside the range of possible values for the problem. For a larger field, we would have to work with pairs of three digits chosen in the same way, again rejecting any numbers outside the possible range of values.

Sampling with and without replacement

A sample may be chosen from a finite population *with replacement* or *without replacement*. For the former (with replacement), having chosen a value to be included in the sample, that value is returned to the population so that it is available again when the next choice for the sample is made. Sampling without replacement means that a value once chosen is not returned to the population and so is not available for choosing when further choices are made. In this case, the size of the population available for choosing reduces by one as each member of the sample is chosen. A finite population in which sampling is with replacement can be treated as an infinite population since, however many values are chosen from it, we can never run out of available values.

Example 1   Each of three girls writes her name on a piece of paper and places the piece in a box. In turn, the girls each draw one piece of paper at random. Given that all the girls have different names, find the probability that none of the girls will draw the piece of paper which has her own name written on it

(a) if each paper is replaced after being drawn,
(b) if each paper is not replaced after being drawn.

(a) Let E be the event that a girl chooses a paper with her own name on it.

$$P(E') = 2/3, \quad P(E) = 1/3, \text{ for every choice.}$$

$$P(E' \cap E' \cap E') = P(E') \times P(E') \times P(E') = \tfrac{2}{3} \times \tfrac{2}{3} \times \tfrac{2}{3} = \tfrac{8}{27}.$$

(b) After the first girl has chosen, there are now only two pieces of paper left in the box (one of which is the name of the girl who chooses first), and, after the second girl has drawn, there is only one piece left for the third girl to take. Hence,

$$P(E' \cap E' \cap E') = \tfrac{2}{3} \times \tfrac{1}{2} \times 1 = \tfrac{1}{3}.$$

Example 2   A population consists of the five digits

$$1 \quad 3 \quad 5 \quad 7 \quad 9.$$

Write down all the possible samples of two digits that could occur if random samples are drawn from this population without replacement.
   Find the mean of each sample and also the mean of the population.
   Find also the mean of the sample means.
   Plot a frequency diagram for the population and for the sample means.

There are $\binom{5}{2}$ possible samples. These are

1 and 3, 1 and 5, 1 and 7, 1 and 9, 3 and 5,
3 and 7, 3 and 9, 5 and 7, 5 and 9, 7 and 9.

Their means are respectively 2, 3, 4, 5, 4, 5, 6, 6, 7, 8.

$$\text{Population mean} = \frac{1 + 3 + 5 + 7 + 9}{5} = 5.$$

$$\text{Mean of sample means} = \frac{2 + 3 + 4 + 4 + 5 + 5 + 6 + 6 + 7 + 8}{10} = 5.$$

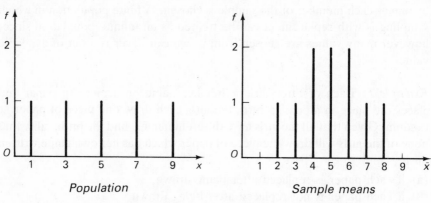

Fig. 2.1

Thus the mean of the sample means is equal to the population mean. As we would expect, we see in Fig. 2.1 that the means of the samples are more closely grouped around the mean, 5, than is the population distribution.

Exercise 2.1

1 A population consists of the five digits 1, 2, 4, 6, 7. Calculate the mean and the standard deviation for this population.

   List all the possible distinct samples of size two which can be drawn, without replacement, from this population.

   Find the mean of each sample, and calculate the mean and the variance of the sample means.

   Sketch the distributions of the population and of the sample means.

2 For a population which consists of the four numbers 1, 3, 6, 10, find the mean and the variance.

   List all the possible samples of size two which can be taken with replacement (that is, for example (6, 6) is a possible sample and (1, 3) and (3, 1) are different samples) from this population.

   Sketch the distribution of the population and of the sample means on the same diagram.

Find the expected value of the sample mean of a sample chosen at random from these samples.

Find also the variance of the distribution of sample means.

2.2 Sampling distributions

Let us look more generally at some sampling distributions. Suppose that we take, with or without replacement, all possible samples of size n which can be drawn from a given population. For each of the samples we could calculate any of the sample statistics of Chapter 1. We might calculate all the sample means, as we did in Example 2 above, or we might, for example, calculate all the sample variances or standard deviations. For each sample statistic we will have a set of values, one from each sample; that is, for each of the statistics that we decide to calculate, we will have a distribution, and this is called the *sampling distribution* of that statistic. The standard deviation of each sampling distribution is called the *standard error* of that statistic.

Example 3 Sketch the sampling distribution of the standard deviations for the data of Example 2 above.

The samples are $(1, 3), (1, 5), (1, 7), (1, 9), (3, 5), (3, 7), (3, 9), (5, 7), (5, 9),$ $(7, 9),$ with means, respectively, $2, 3, 4, 5, 4, 5, 6, 6, 7, 8.$ Using the definition of the standard deviation given in Chapter 1,

$$s = \sqrt{\left(\frac{\sum\limits_{r=1}^{n} (x_r - \bar{x})^2}{n} \right)},$$

we have standard deviations, respectively,

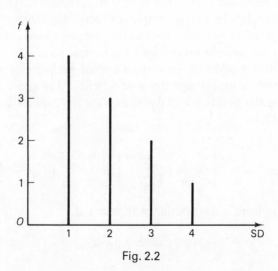

Fig. 2.2

$$\sqrt{\left(\frac{2}{2}\right)} = 1, \ \sqrt{\left(\frac{8}{2}\right)} = 2, \ \sqrt{\left(\frac{18}{2}\right)} = 3, \ \sqrt{\left(\frac{32}{2}\right)} = 4, \ \sqrt{\left(\frac{2}{2}\right)} = 1,$$

$$\sqrt{\left(\frac{8}{2}\right)} = 2, \ \sqrt{\left(\frac{18}{2}\right)} = 3, \ \sqrt{\left(\frac{2}{2}\right)} = 1, \ \sqrt{\left(\frac{8}{2}\right)} = 2, \ \sqrt{\left(\frac{2}{2}\right)} = 1.$$

This sampling distribution is shown in Fig. 2.2.

The sampling distribution of means

In the examples of the previous paragraphs, we have dealt with all possible samples of a given size taken from a finite population of fairly small size. Suppose now that we have an infinite population and, theoretically, we are going to draw all possible samples of size n from this population. Of course, it is not possible to do this in practice, for there are infinitely many such samples. The mean and the variance of a *population* are usually denoted by the symbols μ (mu), for the mean, and σ^2 (sigma squared) for the variance.

Each of the samples will have a mean value, $\bar{x}$, and hence, theoretically, we can sketch the distribution of these means as we did for the finite population case in Example 2 above. Certain properties of this distribution can be deduced from previous working; we would expect the mean of the distribution of sample means to be equal to μ, the mean of the population (as we found in Example 2), and we would expect the sample means to be more closely grouped around the mean μ than is the population. That is, we would expect the standard deviation of the distribution of sample means to be smaller than the standard deviation of the original population.

Also, we would expect the sample means to group themselves more tightly around μ, and therefore the standard deviation of the distribution of sample means to become smaller, as n, the sample size, increases. This latter can be demonstrated in class by taking samples of increasing size and with replacement from a given population, finding the sample means, and plotting the distribution of the sample means for each sample size.

The standard deviation of the distribution of sample means is called *the standard error of the mean*, sometimes written $SE_{\bar{x}}$. The general result can be obtained, using the notation and definitions for the expectation of $\bar{X}$, $E(\bar{X})$, and for $Var(\bar{X})$:

$$E(\bar{X}) = E\left[\frac{\sum_{r=1}^{n} x_r}{n}\right] = \frac{1}{n} E\left[\sum_{r=1}^{n} x_r\right] = \frac{1}{n} \sum_{r=1}^{n} E(x_r) = \frac{1}{n} \sum_{r=1}^{n} \mu,$$

since each x_r is from a distribution with mean μ,

$$\Rightarrow E(\bar{X}) = \frac{1}{n} (n\mu) = \mu.$$

$$\text{Var}(\bar{X}) = \text{Var}\left[\frac{1}{n}\sum_{r=1}^{n}x_r\right] = \frac{1}{n^2}\text{Var}\sum_{r=1}^{n}x_r = \frac{1}{n^2}\sum_{r=1}^{n}\text{Var}(x_r),$$

since $x_1, x_2, \ldots, x_n$, are independent,

$$\Rightarrow \text{Var}(\bar{X}) = \frac{1}{n^2}\sum_{r=1}^{n}\sigma^2 = \frac{n\sigma^2}{n^2} = \frac{\sigma^2}{n}.$$

That is, the standard deviation of the distribution of sample means, $\text{SE}_{\bar{x}}$, is $\sigma/\sqrt{n}$.

Thus the means of all possible samples of size n taken from an infinite population (or from a finite population with replacement) have a distribution whose mean is equal to the mean of the population, μ, and whose standard deviation is equal to the standard deviation of the population divided by $\sqrt{n}$.

We now bring into this work the concept of a normal distribution which is discussed in Chapter 4 of the book *Probability* in this series. A very important theorem, called *the Central Limit theorem*, tells us even more about the distribution of the sample means. It tells us that, if the original population is normally distributed with mean μ and standard deviation σ, that is $X \sim N(\mu, \sigma^2)$, then the distribution of sample means of samples size n is also normally distributed with mean μ and standard deviation $\sigma/\sqrt{n}$, that is $\bar{X} \sim N(\mu, \sigma^2/n)$. If the original population is not normally distributed, the distribution of sample means is approximately normally distributed *provided n is large enough*, the approximate distribution being $N(\mu, \sigma^2/n)$. When the form of the population distribution is asymmetric and therefore far from normal, we shall need a larger sample size n in order to use the approximation than we need when the population distribution is nearly normal. That is, the least n required in order to use the approximation depends on the shape of the population distribution. However, for most practical purposes, $n \geqslant 30$ is acceptable.

The proof of the Central Limit theorem is too difficult to include in this book, but the results can be demonstrated in classwork by taking, with replacement, a number of samples for each of several increasing sample sizes from a known population (for example, the weights of all the pupils in the sixth form of a school), and then plotting the means for each sample size. You will see that, as the sample size increases, the distribution of the means approaches more and more closely the shape of a normal distribution.

Example 4 The mean of a random sample of size 9, taken from an infinite discrete population with mean 40, and variance 16, is denoted by $\bar{x}$. The mean of a random sample of size 18 taken from another infinite discrete population with mean 20, and variance 8, is denoted by $\bar{y}$. Find the expectation and the variance of

(a) $\bar{x} + \bar{y}$, (b) $\bar{x} + 2\bar{y}$, (c) $2\bar{x} - 3\bar{y}$.

In Chapter 2 of the book *Probability* in this series we showed that, if X and Y are two independent discrete random variables, then

$$E(X \pm Y) = E(X) \pm E(Y) \quad \text{and} \quad \text{Var}(X \pm Y) = \text{Var}(X) + \text{Var}(Y).$$

We also proved that

$$E(aX + b) = aE(X) + b \quad \text{and} \quad \text{Var}(aX + b) = a^2\text{Var}(X).$$

Using these results we have

(a) $\bar{x} + \bar{y}$ has mean $40 + 20 = 60$, variance $\dfrac{16}{9} + \dfrac{8}{18} = \dfrac{20}{9}$.

(b) $\bar{x} + 2\bar{y}$ has mean $40 + 40 = 80$, variance $\dfrac{16}{9} + 4 \times \dfrac{8}{18} = \dfrac{32}{9}$.

(c) $2\bar{x} - 3\bar{y}$ has mean $80 - 60 = 20$, variance $4 \times \dfrac{16}{9} + 9 \times \dfrac{8}{18} = \dfrac{100}{9}$.

Example 5 A random sample of twenty-five "1-kg size" packets of porridge oats from a factory production line is taken and weighed. It is known from production records over a long period that the weights of the packets produced by the factory are normally distributed with mean 1·01 kg and standard deviation 0·005 kg. Find the probability that the mean weight of the sample is less than 1·008 kg.

For the population $X \sim N(1 \cdot 01, (0 \cdot 005)^2)$.

The Central Limit theorem tells us that

$$\bar{X} \sim N\left(1 \cdot 01, \frac{(0 \cdot 005)^2}{25}\right).$$

Hence $P(\bar{X} < 1 \cdot 008) = P\left(\bar{Z} < -\dfrac{0 \cdot 002}{0 \cdot 005} \times \sqrt{(25)}\right)'$

$\qquad\qquad\qquad\quad = P(\bar{Z} < -2)$

$\qquad\qquad\qquad\quad = 1 - \Phi(2)$

$\qquad\qquad\qquad\quad = 0 \cdot 0228.$

Example 6 Television tubes produced by a certain company are known to have a mean lifetime of 90 months with standard deviation 8 months. Fifty samples of 100 tubes produced by this company are tested. Estimate, to 2 decimal places, the number of samples which would be expected to have a mean lifetime of more than 91 months.

Let X months be the lifetime of a tube. Then X is distributed with mean 90 months and standard deviation 8 months. Sample size, n, is 100, which is large, hence $\bar{X}$, the mean sample lifetime, is *approximately* normally distributed with mean 90 and standard deviation $8/(\sqrt{100}) = 0 \cdot 8$. That is, $SE_{\bar{x}} = 0 \cdot 8$. Hence,

$$P(\bar{X} \leqslant 91) \approx \Phi(1/0\cdot8) = \Phi(1\cdot25) = 0\cdot89435.$$

Therefore, in 50 samples, we would expect $50 \times 0\cdot89435 = 44\cdot72$ samples to have a mean lifetime less than or equal to 91 months. We would expect approximately 5 samples to have a mean lifetime of more than 91 months.

Example 7 One hundred samples, of constant size n, are taken from a normally distributed population with mean 60 and variance 25. It is found that two of the samples have a mean value greater than 62. Find the probable value of n.

$$X \sim \text{N}(60, 25) \Rightarrow \bar{X} \sim \text{N}\left(60, \frac{25}{n}\right), \text{ where } n \text{ is the sample size.}$$

$$P(\bar{X} > 62) = 1 - P(\bar{X} \leqslant 62) = \frac{1}{50}$$

$$\Rightarrow P(\bar{X} \leqslant 62) = 0\cdot98$$

$$\Rightarrow P\left(\bar{Z} \leqslant \frac{2\sqrt{n}}{5}\right) = 0\cdot98$$

$$\Rightarrow \Phi\left(\frac{2\sqrt{n}}{5}\right) = 0\cdot98$$

$$\Rightarrow \frac{2\sqrt{n}}{5} = 2\cdot054$$

$$\Rightarrow n = \frac{25 \times (2\cdot054)^2}{4} \approx 26\cdot37, \text{ that is, } n = 26.$$

Exercise 2.2

1 For question 2, Exercise 2.1, check that your answers satisfy the Central Limit theorem.
2 The weights (kg) of sacks of sand in a builder's yard are known to be normally distributed with mean 30 and standard deviation 2. Calculate, to three decimal places, the probability that 16 sacks taken at random from the stock in the yard will have a mean weight
   (a) of 31 kg or more,
   (b) lying between 29 and 31·5 kg.
3 In the previous question, find the size of the sample that it is necessary to take if the probability that the sample mean will be within 0·2 of the population mean is at least 0·90.
4 In a distillery it is known that the alcohol content per litre of distillate is distributed with mean 14·2 and standard deviation 0·8155. Given that 400 measurements are made at random during production, find, to three decimal places, the probability that the mean of these measurements will not exceed 14·1.

Miscellaneous Exercise 2

1 Write down the mean and variance of the distribution of means of all possible samples of size n taken from an infinite population with mean μ and variance σ^2.
   Describe the form of this distribution of sample means when

(a) n is large,

(b) the distribution of the original population is normal.

The weights of packets of coffee are normally distributed with a mean of 250 g and a standard deviation of 3 g. Find the probability that the total weight of a random sample of 6 packets will lie between 1488 g and 1506 g.

Estimate, to the nearest gram, the limits within which 95% of the total weights of all possible samples of 25 packets will lie.

2 Fishing floats are being produced in a factory and, over a long period, the population of their diameters is found to have a mean of 45 cm with standard deviation 2 cm. Estimate, to four decimal places, the probability that a random sample of 49 floats will have a mean diameter greater than 45·5 cm.

In 400 random samples each of 49 floats, calculate the expected number of the 400 samples having mean diameter less than 45·5 cm.

3 Records of the diameters of spherical ball-bearings produced on a certain machine show that the diameters are normally distributed with mean 0·821 cm and standard deviation 0·050 cm. Three hundred samples, each consisting of 100 ball bearings, are chosen at random. Calculate the expected number of the 300 samples having a mean diameter less than 0·820 cm.

4 In a test for fading of cloth, 3 tests were carried out in which strips of cloth are exposed to certain amounts of sunlight. In the first test 50 strips of cloth were used, 25 in the second test, and 35 in the third test. The percentage of fading was measured and the results are summarised below.

Sample number	Size of sample	Mean%	Standard deviation %
1	50	2·3	0·25
2	25	2·2	0·30
3	35	2·0	0·10

Find, to two decimal places, the mean percentage fading and the standard deviation for the combined sample of 110 strips of cloth.

5 A normally distributed population of lengths of stag beetles has a mean of 60 mm and a standard deviation of 10 mm. Find the proportion of the population that is less than 65 mm in length.

Find also the probability that a random sample of 16 beetles will have a mean length exceeding 67 mm.

3 Estimation and confidence intervals

3.1 Estimation

In Chapter 2 we stated that often we wish to infer or to predict information about a population by considering a sample (or samples) of collected data consisting of a finite number of observations or measurements from that population. Also, even if the population is large but finite, because of restrictions in time and expense, usually it is not possible to collect all the data for the population, and we certainly cannot collect all the data for an infinite population. We wish to find estimates of *population parameters* (for example, population mean, or variance) from the corresponding *sample statistics* (for example, sample mean, or variance). If the mean of the sampling distribution of a particular statistic is equal to the corresponding population parameter, then we say that the statistic is *an unbiased estimator* of the corresponding population parameter. For example, in Chapter 2, we showed that the mean of the sampling distribution of means is equal to μ, the population mean. Hence, the sample mean, $\bar{x}$, is an unbiased estimator of the population mean μ.

Example 1 A psychologist wishes to estimate the mean reaction-time of car drivers. She takes a random sample of 10 car drivers and finds that their reaction times, in seconds, are as follows:

$$0.11 \quad 0.12 \quad 0.18 \quad 0.16 \quad 0.24 \quad 0.74 \quad 0.43 \quad 0.52 \quad 0.10 \quad 0.27.$$

From these sample data find an unbiased estimate of the mean reaction-time of car drivers.

A good notation to use when estimating parameters is to write a circumflex or 'hat' over the symbol for the parameter to show that it is an *estimated* value only, and not necessarily the true value. Thus we can use $\hat{\mu}$ for an estimated value of the population mean, and $\hat{\sigma}$ for an estimated value of the population standard deviation.

For our sample, the sample mean, $\bar{x}$, is given by

$$\bar{x} = \frac{0.11+0.12+0.18+0.16+0.24+0.74+0.43+0.52+0.10+0.27}{10} \text{ seconds}$$

$$= 0.287 \text{ seconds.}$$

An unbiased estimate of the population mean is, therefore, $\hat{\mu} = 0.287$ seconds.

For an unbiased estimate of the population variance, $\hat{\sigma}^2$, we use $(ns^2)/(n-1)$ where s^2 is the variance of the sample, and n is the size of the sample. Therefore

$$\hat{\sigma}^2 = \frac{\sum_{r=1}^{n} (x_r - \bar{x})^2}{n-1},$$

where the sample is $x_1, x_2, \ldots, x_n$, with mean $\bar{x}$.

We need to show that the sampling distribution of

$$\frac{ns^2}{n-1} = \frac{\sum_{r=1}^{n} (x_r - \bar{x})^2}{n-1}$$

has mean, or expectation, equal to σ^2 if this expression for $\hat{\sigma}^2$ is to fulfil the definition for an unbiased estimator.

First we write

$$\sum_{r=1}^{n} (x_r - \mu)^2 = \sum_{r=1}^{n} [(x_r - \bar{x}) + (\bar{x} - \mu)]^2 \qquad \text{(where } \mu \text{ is the population mean)}$$

$$= \sum_{r=1}^{n} (x_r - \bar{x})^2 + 2(\bar{x} - \mu) \sum_{r=1}^{n} (x_r - \bar{x}) + \sum_{r=1}^{n} (\bar{x} - \mu)^2$$

$$= \sum_{r=1}^{n} (x_r - \bar{x})^2 + (\bar{x} - \mu)^2 \sum_{r=1}^{n} 1,$$

since, by definition of $\bar{x}$, $\sum_{r=1}^{n} (x_r - \bar{x}) = 0$. Therefore

$$\sum_{r=1}^{n} (x_r - \mu)^2 = \sum_{r=1}^{n} (x_r - \bar{x})^2 + n(\bar{x} - \mu)^2$$

$$\Rightarrow \sum_{r=1}^{n} (x_r - \bar{x})^2 = \sum_{r=1}^{n} (x_r - \mu)^2 - n(\bar{x} - \mu)^2.$$

Hence

$$\frac{1}{n-1} \mathrm{E} \sum_{r=1}^{n} (x_r - \bar{x})^2 = \frac{1}{n-1} \mathrm{E} \sum_{r=1}^{n} (x_r - \mu)^2 - \frac{n}{n-1} \mathrm{E}(\bar{x} - \mu)^2$$

$$= \frac{1}{n-1} \sum_{r=1}^{n} \mathrm{E}(x_r - \mu)^2 - \frac{n}{n-1} \times \frac{\sigma^2}{n},$$

since the variance of the distribution of the sample means, $\mathrm{E}(\bar{x} - \mu)^2$, is equal to σ^2/n (cf. § 2.2). Also, $\mathrm{E}(x_r - \mu)^2 = \sigma^2$, the population variance, and hence

$$\frac{1}{n-1} \text{E} \sum_{r=1}^{n} (x_r - \bar{x})^2 = \frac{1}{n-1} \sum_{r=1}^{n} \sigma^2 - \frac{\sigma^2}{n-1}$$

$$= \frac{n}{n-1} \sigma^2 - \frac{1}{n-1}\sigma^2 = \sigma^2,$$

as required. Thus $\frac{n}{n-1} s^2$ is an unbiased estimator of the population variance.

Example 2 For the data of Example 1, find, to three decimal places, an unbiased estimate of the variance of the population.

We had $\bar{x} = 0\cdot287$ seconds. $\sum_{r=1}^{n} (x_r - \bar{x})^2$ for the ten reaction times is equal to

$[0\cdot0313 + 0\cdot0279 + 0\cdot0114 + 0\cdot0161 + 0\cdot0022 + 0\cdot2052 + 0\cdot0204 + 0\cdot0543 + 0\cdot0350 + 0\cdot0003]$ (seconds)^2 = $0\cdot4041$ (seconds)^2.

Also $n = 10$. Therefore

$$\hat{\sigma}^2 = \frac{0\cdot4041}{(10 - 1)} \text{ (seconds)}^2 = 0\cdot045 \text{ (seconds)}^2.$$

Estimates of population parameters such as we have found in these two examples, i.e. an estimate of the population mean $\hat{\mu} = 0\cdot287$ seconds in the first example, and an estimate of the population variance $\hat{\sigma}^2 = 0\cdot045$ (seconds)^2 in the second example, are called *point estimates* of the parameter concerned.

Exercise 3.1

1 A random sample of five readings

$$1\cdot40 \quad 1\cdot39 \quad 1\cdot34 \quad 1\cdot32 \quad 1\cdot35$$

is taken from a population. Find unbiased estimates for the mean and the variance of the population.

2 A sample of 8 metal rods chosen at random from the production of a machine is found to have mean length $1\cdot01$ m and variance $0\cdot007$ m^2. Find unbiased estimates of the mean and the variance of the population of metal rods produced by the machine.

3 A random sample of 10 light bulbs produced by a company showed a mean lifetime of 620 hours with standard deviation of 21 hours. Estimate the mean and the standard deviation of the population of light bulbs produced by this company.

3.2 Confidence limits for a population mean

Obviously, for different samples from the same population we generally obtain different estimates, $\hat{\mu}$ and $\hat{\sigma}^2$, of μ and σ^2. It would be better if we could give an interval within which we could say, with a stated degree of certainty, that a population parameter will lie. An estimate of a population parameter given in this way is called an *interval estimate* of the parameter.

The Central Limit theorem states that the distribution of means of samples of size n is normally distributed if the population is normally distributed, and is approximately normally distributed when n is large ($\geqslant 30$), even if the population is not normally distributed. We know also that the mean and variance of the distribution of sample means are μ and σ^2/n, respectively. Thus in these cases we can, using normal area tables, give limits, symmetrical about the mean, within which we expect a stated percentage of sample means to lie.

Example 3 Find the limits between which we would expect 98% of the sample means to lie.

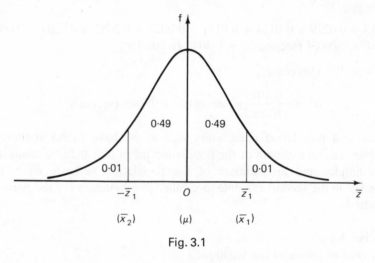

Fig. 3.1

Figure 3.1 shows the distribution, or the approximate distribution, of the sample means.

$$\Phi(\bar{z}_1) = 0\cdot99 \Rightarrow \bar{z}_1 = 2\cdot237 = \frac{\bar{x}_1 - \mu}{\sigma/\sqrt{n}}$$

$$\Rightarrow \bar{x}_1 = \mu + 2\cdot327 \, \sigma/\sqrt{n}.$$

Hence $\qquad \bar{x}_2 = \mu - 2\cdot327 \, \sigma/\sqrt{n}.$

That is, 98% of the means lie within the interval $\mu \pm 2\cdot327 \, \sigma/\sqrt{n}$.
Similarly, it can be shown that

$$99\% \text{ lie in the interval } \mu \pm 2\cdot576 \, \sigma/\sqrt{n},$$
$$95\% \text{ lie in the interval } \mu \pm 1\cdot960 \, \sigma/\sqrt{n}, \text{ etc.}$$

We saw in §3.1 how we can find unbiased estimates, $\hat{\mu}$ and $\hat{\sigma}^2$, of μ and σ^2, the population mean and variance, when these are not known. These estimates were derived from the statistics of a random sample taken from the population. Provided $n \geqslant 30$, use of the unbiased estimate of the population variance leads to quite satisfactory results when we are finding *a confidence interval for the population mean*, but if $n < 30$ then the approximation is generally very poor. We use the same method that we employed in the last example except that now we have an unknown population mean which has been estimated by $\hat{\mu} = \bar{x}$, the sample mean. We will consider three cases.

Case 1 Given that the population is normally distributed and that σ^2, the population variance, is known, we use the fact that the sampling distribution of means is also normally distributed for all values of n with variance σ^2/n. We can then set up $k\%$ *confidence limits* (where k is a given constant) for the unknown population mean.

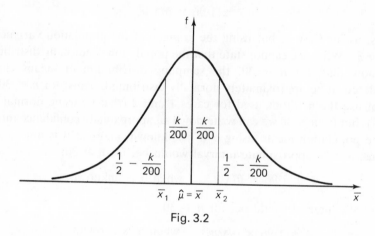

Fig. 3.2

The confidence limits, $\bar{x}_1$ and $\bar{x}_2$, are taken to be symmetrical about the estimated mean $\bar{x}$, so that $k\%$ of the distribution lies between them ($\bar{x}_1$ and $\bar{x}_2$), and $(100 - k)\%$ lies outside them, distributed symmetrically on each side as shown in Fig. 3.2. We then look up in a normal area table the value, $\bar{z}_1$, of $\bar{z} = \dfrac{\bar{x}_2 - \bar{x}}{\sigma/\sqrt{n}}$, for which

$$\Phi(\bar{z}_1) = \frac{k}{100} + \frac{1}{2} - \frac{k}{200} = \frac{1}{2} + \frac{k}{200}.$$

The $k\%$ confidence limits for μ are then $-\bar{z}_1$ and $+\bar{z}_1$ or $\bar{x}_1 = \bar{x} - \bar{z}_1 \times \sigma/\sqrt{n}$ and $\bar{x}_2 = \bar{x} + \bar{z}_1 \times \sigma/\sqrt{n}$.

For 95% confidence limits we have

$$\Phi(\bar{z}_1) = 0 \cdot 975 \Rightarrow \bar{z}_1 = \pm 1 \cdot 96,$$

and hence the limits are $\bar{x} \pm 1 \cdot 96 \times \sigma/\sqrt{n}$.

Similarly, 98% confidence limits for the population mean are $\bar{x} \pm 2 \cdot 327 \times \sigma/\sqrt{n}$.

We can write down other confidence limits by looking up the appropriate $\bar{z}$ values in a normal area table. The interval between the two limits is called *the confidence interval*.

Case 2 In this case we are given that the population is normally distributed, but that the population variance is unknown; here again we know that the sampling distribution of means is normally distributed. We can find an unbiased estimate $\hat{\sigma}^2$ of the population variance, but, unless $n \geqslant 30$, this estimate is generally not good enough for us to use in forming a confidence interval for the population mean by the use of normal area tables, as we did in Case 1.

Thus, when $n \geqslant 30$, an approximate 95% confidence interval for the population mean can be written as

$$\bar{x} \pm 1.96 \times \hat{\sigma}/\sqrt{n},$$

that is, as for Case 1, but using the *estimate* of the population variance.

Case 3 When we cannot state that the population is normally distributed, we know that, for $n \geqslant 30$, the sampling distribution of means can be considered to be approximately normally distributed. Thus, for $n < 30$, we cannot use the methods used for cases 1 and 2 (that is, using normal area tables), but for $n \geqslant 30$ we can write down an approximate confidence interval for the population mean, using σ if it is known, and $\hat{\sigma}$ if it is not.

Thus, a 95% approximate interval would be, for $n \geqslant 30$,

$$\bar{x} \pm 1.96 \times \hat{\sigma}/\sqrt{n}, \qquad \text{when } \sigma \text{ is unknown,}$$

and a 95% interval would be, for $n \geqslant 30$,

$$\bar{x} \pm 1.96 \times \sigma/\sqrt{n}, \qquad \text{when } \sigma \text{ is known,}$$

We discuss how we can deal with some of the cases for which $n < 30$, that is, for some small samples, in Chapter 7 when we consider the t-test.

Example 4 Measurements of the lengths of a random sample of 100 bolts produced by a machine during one day showed that they had a mean length of 2·2 cm with standard deviation 0·05 cm. Find (a) a 90%, (b) a 99%, confidence interval for the mean length of a bolt produced by the machine.

This is Case 3, with $n = 100$. We have only the sample variance $(0·05)^2 = s^2$

$$\Rightarrow \hat{\sigma}^2 = \frac{100}{99} (0·05)^2 \text{ (cm)}^2 = 0·00253 \text{ (cm)}^2$$

$$\Rightarrow \hat{\sigma} = 0·0503 \text{ cm.}$$

(This shows that for large samples, that is for large values of n, $\hat{\sigma} \approx s$, as we would expect from the result $\hat{\sigma}^2 = \frac{n}{n-1}s^2$.)

(a) 90% confidence limits are

$$\left(2 \cdot 2 \pm 1 \cdot 645 \times \frac{0 \cdot 0503}{\sqrt{100}}\right) \text{cm}$$

$$= (2 \cdot 2 \pm 0 \cdot 0083) \text{ cm}$$

$\Rightarrow$ a 90% confidence interval is $(2 \cdot 192 \text{ cm to } 2 \cdot 208 \text{ cm})$.

(b) 99% confidence limits are

$$\left(2 \cdot 2 \pm 2 \cdot 576 \times \frac{0 \cdot 0503}{\sqrt{100}}\right) \text{cm}$$

$$= (2 \cdot 2 \pm 0 \cdot 012\,96) \text{ cm}$$

$\Rightarrow$ a 99% confidence interval is $(2 \cdot 187 \text{ cm to } 2 \cdot 213 \text{ cm})$.

In words, this tells us that we can say with 90% certainty that the population mean lies between 2·192 cm and 2·208 cm, and with 99% certainty that it lies between 2·187 cm and 2·213 cm.

Example 5 Children entering nursery school are given a test. It is known that the scores for children taking this test are normally distributed with variance 16. A random sample of 10 children entering at the start of one year are found to have a mean score of 90. On the basis of this result, find a 95% confidence interval for the mean score of all children taking the test.

This is Case 1 where we know that the population is normally distributed and that the population variance is 16, so it does not matter that the sample is small in size.

We have $\bar{x} = 90$, $n = 10$, $\sigma^2 = 16$, so a 95% confidence interval for the population mean is given by

$$\left(90 - 1 \cdot 96 \times \frac{4}{\sqrt{10}}\right) \text{ to } \left(90 + 1 \cdot 96 \times \frac{4}{\sqrt{10}}\right),$$

that is, 87·52 to 92·48.

Example 6 The standard deviation of the lifetimes of a certain make of car battery is estimated to be 220 hours. Find the size, n, of the smallest random sample of these batteries which we should take in order that we can be 99% confident that the error in the estimated mean lifetime will not exceed 40 hours.

99% confidence limits for the population mean are $\bar{x} \pm 2 \cdot 576 \times \hat{\sigma}/\sqrt{n}$, provided $n \geqslant 30$. Here $\hat{\sigma} = 220$ and n is the sample size. We require

$$\frac{2 \cdot 576 \times 220}{\sqrt{n}} < 40$$

$\Rightarrow \sqrt{n} > 14 \cdot 168$

$\Rightarrow n > 200 \cdot 732$

$\Rightarrow n = 201$ is the smallest sample size.

Exercise 3.2

1 Find the probability of obtaining a sample with mean 8·4 when a random sample of size 16 is taken from a normal population with mean 8·5 and variance 0·4.

2 A random sample taken from a normal population with standard deviation 2 consisted of the readings

$$4, 5, 2, 4, 8, 9, 3, 15.$$

Find, to three decimal places, a 90% confidence interval for the mean of the population.

3 A random sample of 100 batteries, each of nominal voltage 1·5 volts, was taken from a large production and the actual voltages measured. The results were as follows:

Voltage (mid-interval)	1·35	1·40	1·45	1·50	1·55	1·60	1·65
Number of batteries	5	10	24	32	20	8	1

Calculate the mean voltage and the standard deviation of this sample of batteries.

Estimate, to four decimal places, the standard error of the mean, and use it to estimate, to three decimal places, 95% confidence limits for the mean voltage of batteries being produced.

4 A survey of 1000 women in a certain region who do part-time work showed their average weekly wage to be £28·40 with standard deviation £4·80. Assuming that this is a random sample from the population of women with part-time earnings in that region, estimate, to two decimal places, 98% confidence limits for the mean part-time wage of women in that region.

Given the same mean wage and standard deviation, estimate how large the sample would have to be if the 98% confidence interval for the population mean was £(28·40 ± 0·20).

Miscellaneous Exercise 3

1 For a normally distributed population of heights of men with mean 67 inches and standard deviation 2·5 inches, find the probability that a sample of 100 such heights should have a mean differing by more than 0·5 inches from the population mean.

2 In a test on the breaking strengths of a particular type of rope being produced, 36 pieces of rope chosen at random are tested. The breaking strength, x, of each piece of rope is recorded, with the results $\sum_{r=1}^{36} x_r = 342$, $\sum_{r=1}^{36} x_r^2 = 3340$. Assuming that neither the mean nor the standard deviation of the population of breaking strengths of ropes being produced is known, find estimates of the population mean and variance.

3 A random sample of size n is taken from a normal distribution with mean μ and standard deviation 4. Find the least value of n in order that the probability that the difference between the sample mean and the population mean does not exceed 0·5 is at least 0·95.

4 The breaking strengths, x kN, of 15 lengths of heavy nylon thread were measured. A summary of the results is shown.

$$\Sigma x = 420, \qquad \Sigma x^2 = 12016.$$

Estimate the mean and the variance of the population from which these experimental results arose.

Assuming that this population was normal with variance 4 $(kN)^2$, find, to two decimal places, 90% confidence limits for the population mean.

Estimate the number of measurements needed so that the 90% confidence interval for the mean covers a range of values of not more than 0·5 kN.

5 Measurements $x_1, x_2, \ldots, x_{150}$ of the value of a variable x made on a random sample of 150 articles gave $\sum_{r=1}^{150} x_r = 1365$, $\sum_{r=1}^{150} x_r^2 = 16380$. Estimate, to three decimal places, the standard error of the mean of x.

Assuming that the population variance is known to be 26·567, find the number of measurements necessary for the mean of the distribution to be known, at the 95% confidence level, to within 0·2 of its value.

6 Given that a sample of 9 values from a normal population of known variance led to a 95% confidence interval for the population mean of 8·5 to 11·7, calculate the width of a 95% confidence interval which would arise from a sample of 100 values from the same population.

7 The amount of a chemical in effluent from a power station is measured frequently. For 400 measurements during one week, the amount of chemical per litre is shown below.

Chemical content	12	13	14	15	16
Frequency	5	52	235	74	34

Calculate the mean chemical content, and estimate the standard error of the mean.

Obtain 95% confidence limits for the mean chemical content in the effluent.

Given that a further 400 measurements are to be made during that week, find an approximate value for the probability that the mean of these measurements will not exceed 14·1.

8 The error made when using a certain instrument to measure the lengths of a certain type of earwig is known to be normally distributed with mean 0 and standard deviation 1 mm. Calculate the probability that the error made when the instrument is used once is in absolute magnitude less than 0·4 mm.

Given that one earwig is measured 9 times with the instrument, find the probability that the mean of the 9 readings will be within 0·5 mm of the true length of the earwig.

9 Write down the mean and the variance of the distribution of means of all possible samples of size n taken from an infinite population with mean μ and variance σ^2. Describe the form of this distribution
(a) when the population is normal,
(b) when the population is not quite normal and n is large.
One hundred measurements of a wavelength give a mean of 411·63 nm with unbiased estimate of the population variance 0·764 $(nm)^2$. Estimate a 99% confidence interval for the true mean wavelength.

10 A normally distributed population of hamster body weights has mean 62·5 g with standard deviation 12·0 g.
(a) Find the proportion of the population with weight greater than 77·0 g.

(b) Find the probability of choosing at random from this population a body weight between 48·6 g and 77·0 g.

(c) Find the standard deviation of all possible sample means of samples of size 10 which can be drawn from this population.

(d) Find the probability of the mean of a sample of size 10 lying between 58·0 g and 60·0 g.

4 Significance testing

4.1 Introduction

At the start of Chapter 2 we said that the managers of a bus company may wish to discover the average journey time of their passengers, and to this end they could take a random sample of their passengers and record the journey times of this sample. As we saw in Chapter 3, they could then estimate the mean journey time of the population of all the passenger journey times. However, the company managers may wish to do more than this. They may wish to make further decisions regarding the population on the basis of the sample information. This could well affect the design of new buses, or the method of fare collection, for example.

A drug company may wish to decide whether a new drug really is effective in curing a particular disease. The drug would be tested on a random sample of patients suffering from the disease and there would also be a 'control group' of sufferers who think they are taking the drug but are, in fact, taking sugar pills (since mental approach may often affect cure). A comparison of the results of these two sets of patients could help the company to make a decision about the effectiveness of the drug in curing the disease.

Decisions such as those arising from these two examples are called *statistical decisions*. To reach such a decision, we start by making assumptions about the population with which we are dealing. We call these assumptions *statistical hypotheses*, and they are usually assumptions about parameters of the probability distribution of the population. We start by assuming that any differences between our observed results and the population of results are due simply to *sampling*, or *chance*, fluctuations. This assumption is called the *null hypothesis* and is denoted by H_0; it is an assumption that there is *no difference* between the population from which the sample results come and the population with which we are concerned. We work on the assumption that the null hypothesis, H_0, is true, and, if we find that the observed results in our random sample have only a small probability of occurring under H_0, then we must reject the idea that this is a random sample from the population under consideration. That is, we must reject H_0 on the basis of our sample evidence. If we reject H_0, then we must have *an alternative hypothesis*, H_1, which we can automatically accept. An alternative hypothesis will usually be one of two kinds; say, for example, that our null hypothesis is that the mean μ_1 of the population from which the sample comes is equal to the mean μ_2 of the population under consideration, that is, $H_0: \mu_1 = \mu_2$. Then the alternative hypothesis could be

either $H_1: \mu_1 \neq \mu_2$, the two population means are different,

or $H_1: \mu_1 > \mu_2$, the mean of the population from which the sample comes is greater than that of the population under consideration,

or $H_1: \mu_1 < \mu_2$, the mean of the population from which the sample comes is less than that of the population under consideration.

The choice of H_1 is entirely dependent on the question being asked and answered. We are testing the null hypothesis H_0; if H_1 is of the first of these three possible forms, where it is concerned with *any* change in the parameter being tested, we say that we are performing a *two-tailed* (or *two-sided*) test. If H_1 is either of the latter two forms, where it is concerned only with strictly an increase or with strictly a decrease in the parameter being tested, we say we are performing a *one-tailed* (or *one-sided*) test.

A *Type 1 error* is made when we reject a hypothesis which should have been accepted.

A *Type 2 error* is made when we accept a hypothesis which should have been rejected.

Both types of error will decrease as the sample size increases; that is, the larger our sample of results, the less likely we are to come to a wrong decision.

We said that we find the probability of obtaining the sample results as a random sample under the null hypothesis H_0, and if this probability is too small then we reject H_0 as being false. The level at which we consider the probability as being sufficiently small for us to reject H_0 will vary according to the importance of the results of the work we are carrying out. In the two examples described at the beginning of this chapter, the experiment to test the efficacy of a new drug has results which are more vital to peoples' well-being than work done on the lengths of passengers' bus journeys, and so we must place more rigid levels on our decision making in the former case. The probability level which we take and which is the maximum probability with which we would be willing to make a Type 1 error, is called the *level of significance* or *significance level* of our test. Many questions on signficance testing in A-level papers ask you to state the significance level which you are using in drawing your conclusions. This might be, for example, 10%, 5%, $2\frac{1}{2}\%$ or 1%. Other questions will tell you to take a particular significance level, usually 5%. A 5% significance level means that, if the probability of getting the sample results on the basis of chance in random sampling under the assumption H_0 is less than 0·05, then we reject H_0 and accept H_1. We say that we are 'rejecting H_0 using a 5% significance level', which means that the probability of our being wrong in rejecting H_0 is only 0·05. For this significance level, the differences between the sample results and those expected by random sampling under H_0 are then said to be *significant*. If the probability exceeds 0·05, the differences are *not significant* and we cannot reject H_0 at the 5% level.

4.2 Significance testing of a population mean using large samples

For the standard normal distribution, N(0, 1), we have, from normal area tables,

$$\Phi(1·960) = 0·975,$$

that is, 95% of the distribution lies within the interval $z = \pm 1·960$. It follows that, for a normal distribution $N(\mu, \sigma^2)$, 95% of the distribution lies within the interval $x = \mu \pm 1·960\ \sigma$. Sometimes we round off the number 1·960 to become 2 and then the '2-σ rule' states that 95% of the readings of a normal distribution lie within approximately two standard deviations of the mean.

We know from Chapter 2 that the sampling distribution of means of samples of size n, *when n is large*, from any distribution with mean μ, variance σ^2, is *approximately* normally distributed $N(\mu, \sigma^2/n)$. It follows that, in either of these circumstances, 95% of the means of samples of size n would be expected to lie within $1·960\ \sigma/\sqrt{n}$ either side of the mean μ. We could, in the same way, write down other confidence intervals, for example, 90%, $97\frac{1}{2}\%$, and 99%, corresponding to probabilities of 10%, $2\frac{1}{2}\%$ and 1%, respectively, of obtaining a sample mean from the population lying outside those intervals (sometimes called the *region of rejection or region of significance*). Thus we could perform a *significance test* in which we could find the probability of obtaining a given sample mean under the null hypothesis that the sample comes from a population of stated mean value.

Example 1　Customer records over a long period have shown that the length of time a certain make of car runs before the first major breakdown is normally distributed with mean 26 months and standard deviation 5 months. A customer complains that his car has had a major breakdown after 15 months. Investigate whether this customer has grounds for his complaint.

Let the time before the first major breakdown be T months. Then $T \approx N(26, 5^2)$. Here we have a single reading of 15 from the T population. Assuming that this reading does come from the T population (this is H_0), we find that

$$P(T \leqslant 15) = \Phi(-11/5) = 1 - \Phi(2·2) = 0·0139.$$

Our test is only *one-tailed* since we are concerned only with T being *as low as or lower than 15*. We will use a 5% significance level. There is only a probability of 0·0139 (approximately 1·4%) of picking a car with a 'first major breakdown time' as low as or lower than this by random sampling from the population of cars being produced. Hence, we reject, at the 5% significance level, the null hypothesis that this is a random sample from the given population, and we decide that the customer does appear to have grounds for complaint.

Had we decided to use a 1% significance level, then our decision would have been different; we cannot then reject the hypothesis that this is a random sample from the population of cars produced, and the customer would appear to have little ground for complaint. A 5% significance level would, generally, be a more reasonable one to use in a problem of this kind.

Example 2   A random sample of size 200 is taken from a population which is known to be normally distributed. From the sample data, it is found that the sample mean is 143 and an estimate of the population variance is 400. Test
(a) at the 5% significance level,
(b) at the $2\frac{1}{2}$% significance level,
the hypothesis that the population mean is equal to 146 against the hypothesis that the population mean is not equal to 146.

$H_0: \mu$ (the population mean) $= 146$.
$H_1: \mu \neq 146$, that is, a two-tailed test.

We test whether a sample of size 200 with $\bar{x} = 143$ and $\hat{\sigma} = 20$ can be considered as coming from from a distribution with mean 146. Since the population is distributed $N(\mu, \sigma^2)$, the sampling distribution of means is distributed $N(\mu, \sigma^2/n)$. We do not know σ, but because n is large ($= 200$) we can use the estimate, $\hat{\sigma} = 20$, instead of σ to find

$$\bar{z} = \frac{143 - 146}{20} \times \sqrt{200} = -2 \cdot 121.$$

(a) The 95% critical limits, two-tailed, are $\bar{z} = \pm 1 \cdot 960$; that is, $P(\bar{z} \geqslant 1 \cdot 96$ or $\leqslant -1 \cdot 96) = 0 \cdot 05$.
$\bar{z} = -2 \cdot 121$ lies within this region of rejection (shaded in Fig. 4.1). Hence, at the 5% level, we reject $H_0: \mu = 146$.

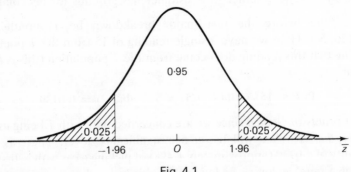

Fig. 4.1

(b) Similarly, the $97\frac{1}{2}\%$ critical limits, two-tailed, are $\bar{z} = \pm\ 2\cdot2418$. $\bar{z} = -2\cdot121$ lies within these limits, in the region of acceptance. Hence, at the $2\frac{1}{2}\%$ significance level, we cannot reject $H_0: \mu = 146$.

Alternatively, we could say, using a two-tailed test,

$$P(\bar{z} \geqslant 2\cdot121 \text{ or } \leqslant -2\cdot121) = 2P(\bar{z} \geqslant 2\cdot121) \text{ by symmetry,}$$
$$= 2 \times 0\cdot01696 = 0\cdot0339,$$

or approximately $3\cdot4\%$.

(a) $P < 5\%$, therefore we reject H_0 at the 5% level,
(b) $P > 2\frac{1}{2}\%$, therefore we cannot reject H_0 at the $2\frac{1}{2}\%$ level.

Example 3 Production records kept over a long period in a factory for a machine producing sewing thread have shown that the breaking strength of the thread is distributed with mean $10\cdot22$ units and standard deviation $1\cdot54$ units. On a particular day, a random sample is taken of 50 threads produced that day, and these are found to have mean breaking strength of $9\cdot80$ units. Investigate, at the 5% significance level, whether or not the factory manager can conclude that the machine is producing inferior thread that day.

Here we do not know the form of the distribution of the population of breaking strengths, but we know that its mean is $10\cdot22$ and its standard deviation is $1\cdot54$. Since $n = 50$, we can say that the sampling distribution of means is *approximately* normally distributed $N\left(10\cdot22, \dfrac{(1\cdot54)^2}{50}\right)$.

$H_0: \mu = 10\cdot22$ (that is, the population from which the sample comes has mean equal to $10\cdot22$),
$H_1: \mu < 10\cdot22$ (one-tailed), since we are concerned only with the mean breaking strength being lowered.

$$P(\bar{x} \leqslant 9\cdot80) = P\left(\bar{z} \leqslant \frac{9\cdot80 - 10\cdot22}{1\cdot54/\sqrt{50}}\right)$$
$$= P(\bar{z} \leqslant -1\cdot9285)$$
$$= 0\cdot0269.$$

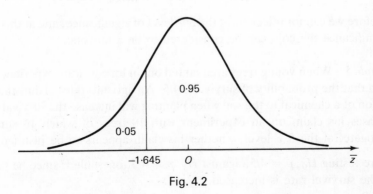

Fig. 4.2

The 5% critical limit, one-tailed, shown in Fig. 4.2, is $\bar{z} = -1.645$. We reject H_0 at the 5% level of significance. It does appear that thread produced that day is weaker.

Exercise 4.2

1 A potato grower states that the mean weight of a sack of his potatoes is 20·0 kg. The weights in kilograms, of a sample of 10 of his sacks are found to be

$$21·2 \quad 21·1 \quad 20·6 \quad 20·1 \quad 19·5 \quad 19·3 \quad 20·9 \quad 20·4 \quad 20·2 \quad 19·7.$$

Find estimates of the mean weight, μ, and of the standard deviation, σ, of a sack of his potatoes.

Assuming that the true value of σ is equal to this estimated value and that the weights of the sacks of potatoes are normally distributed, test, at the 5% significance level, the grower's claim that his sacks of potatoes have a mean weight of 20·0 kg.

2 The breaking strength of a type of cord is known to have a normal distribution with mean μ and standard deviation 1·4 units. A random sample of 49 newly produced lengths of the cord are found to have a mean breaking strength of 9·2 units. Test (a) at the 5% significance level, (b) at the 1% significance level, the null hypothesis that $\mu = 9·6$ units against the alternative hypothesis $\mu < 9·6$ units.

4.3 Significance testing of a binomial parameter

We can use the ideas of §§ 4.1 and 4.2 to test a binomial probability, p.

Example 4 A coin is tossed 8 times. It shows tails seven times. Test whether this result leads you to conclude that the coin is fair.

Here we are testing $H_0: p = \frac{1}{2}$, that is, head and tail have an equal chance of occurring, against $H_1: p \neq \frac{1}{2}$ (two-tailed), that is, the coin is biased. The number of tails, X, is distributed $B(8, p)$.

Under $H_0: p = \frac{1}{2}$, and using a two-tailed test, we have

$$P(X \geqslant 7 \text{ or } \leqslant 1) = 2P(X = 7 \text{ or } 8) \text{ by symmetry when } p = \tfrac{1}{2},$$
$$= 2[8(\tfrac{1}{2})^8 + (\tfrac{1}{2})^8] = 0·0703.$$

Therefore we cannot reject H_0 at the 5% level of significance, and at this level of significance the coin can be considered to be a fair one.

Example 5 When young trees are planted out in forests, long experience has shown that the probability of survival is 4/5. An agriculturalist claims that the addition of a chemical to the soil when planting will increase the survival rate. He bases his claim on an experiment with 18 trees of which 16 survive. Investigate, at the 5% level, whether his claim appears to be justified.

We are testing $H_0: p = 4/5$, against $H_1: p > 4/5$, one-tailed, since he claims that the survival rate is increased.

Let the number of trees surviving be X, then $X \sim B(18, p)$.
Under $H_0: p = 4/5$ (and hence $q = 1/5$),

$$P(X \geqslant 16) = 153\left(\frac{4}{5}\right)^{16} \cdot \left(\frac{1}{5}\right)^2 + 18 \left(\frac{4}{5}\right)^{17} \cdot \left(\frac{1}{5}\right) + \left(\frac{4}{5}\right)^{18}$$
$$= 0 \cdot 271.$$

Hence we cannot reject H_0 at the 5% significance level; the agriculturalist's claim does not appear to be justified.

Here, since $P \approx 27\%$, we could not reject H_0 at the 10% significance level. Had 17 of the 18 trees survived, then

$$P(X \geqslant 17) = \left(\frac{4}{5}\right)^{17} \cdot \left[\frac{18}{5} + \frac{4}{5}\right] = 0 \cdot 0991.$$

We could still not reject H_0 at the 5% level, but we could just reject it at the 10% level, although the result is too close to 10% to make a really satisfactory decision at this level.

In our work on probability, we stated that, when n, the sample size, is large, the finding of binomial probabilities can be tedious, especially when no suitable tables are available. However, we also stated that, for large n, and p not too close to 0 or 1, the normal distribution $N(np, npq)$, where $q = 1 - p$, can be used *as an approximation* to the binomial distribution $B(n, p)$. A common 'rule of thumb' for the conditions on n and p is to use the approximation only when np and nq both exceed 5. Also, we must remember that, when we make this approximation, we are approximating a discrete variable by a continuous one and so we must make a *continuity correction*. We consider each discrete point as representing a corresponding range. For example, the integers 2 and 5 are considered as the corresponding ranges $1\frac{1}{2}$ to $2\frac{1}{2}$ and $4\frac{1}{2}$ to $5\frac{1}{2}$ respectively. Then $P(X \leqslant 2)$ would become $P(X < 2 \cdot 5)$, $P(X > 5)$ would become $P(X \geqslant 5 \cdot 5)$ and $P(4 \leqslant X \leqslant 7)$ would become $P(3 \cdot 5 < X < 7 \cdot 5)$ in the approximate normal distribution. Hence, for large samples, we can use the corresponding normal approximation when performing a significance test on a binomial parameter.

Example 6 A nasal spray is claimed to be 80% effective in relieving asthma attacks. A sample of 300 asthma sufferers used the spray during an attack, and of these 220 said that their symptoms were relieved. Investigate whether these results uphold the claim of 80% effectiveness.

Let p be the probability that the symptoms are relieved. We are testing $H_0: p = 0 \cdot 8$, that is, the claim is true, against $H_1: p < 0 \cdot 8$, that is, the claim is false. A one-tailed test is used since we are interested only in whether the percentage of sufferers relieved is lower than 80%.

We take a 5% significance level; that is, the critical limit for z, one-tailed, is $z = -1 \cdot 645$.

The normal approximation to $B(300, p)$ is $N(300p, 300p(1-p))$. Under the null hypothesis $H_0 : p = 0.8$, that is $N(240, 48)$.

$$P(X \leqslant 220) = P(X < 220.5) \text{ using a continuity correction,}$$
$$= P\left(z < \frac{220.5 - 240}{\sqrt{48}}\right)$$
$$= P(z < -2.815)$$
$$= 0.0024.$$

The value $z = -2.815$ is within the region of rejection of H_0, since $-2.815 < -1.645$. We see that the probability of only 220 sufferers out of 300 having their symptoms relieved if the claim is true is as low as 0.2%. We reject strongly the claim that the spray is 80% effective.

Example 7 Research theory predicts that when mothers who smoke heavily give birth, their babies are more likely than not to have breathing difficulties. A sample of 60 babies born to mothers who smoke heavily is found to include 38 babies with breathing difficulties at birth. Investigate at (a) the 5% significance level, (b) the 1% significance level, whether this confirms the theoretical prediction.

The null hypothesis is that babies are equally likely to have or not to have breathing difficulties at birth $\Rightarrow H_0 : p = \frac{1}{2}, H_1 : p > \frac{1}{2}$, one-tailed since we are concerned only with more than half of them having breathing difficulties.

$n = 60, p = \frac{1}{2}, q = \frac{1}{2}, np = 30, npq = 15$. Thus $B(60, \frac{1}{2})$ can be approximated by $N(30, 15)$. Then

$$P(X > 37.5) = P\left(z > \frac{7.5}{\sqrt{15}}\right)$$
$$= P(z > 1.936)$$
$$= 0.026, \text{ i.e. about 3\%.}$$

(a) For a 5% significance level, the critical value of z, one-tailed, is $z = 1.645$.
(b) For a 1% significance level, the critical value of z, one-tailed, is $z = 2.33$.

Or, we can see that the probability is approximately 3%, that is, less than 5% and greater than 1%. Thus, we can say that the result is significant at the 5% level, and we reject H_0, but that it is not significant at the 1% level and then we cannot reject H_0. Hence, at the 5% level, the results do confirm the theory, since at this level we reject H_0 in favour of H_1, but at the 1% level the theory is rejected and H_0 accepted.

Summary
You will notice that there is a basic pattern to all the examples worked in this chapter. First we write down the null hypothesis H_0, and then we decide on

and write down the alternative hypothesis H_1, which will be either one-tailed or two-tailed depending on the question being answered. We then decide on a significance level (probably 5%), if we are not told in the problem what level to take. *Only then* can we start to do the calculations, using the assumption H_0 for the test. Finally we write down our conclusions so that they can be understood by other readers.

Exercise 4.3

1 Six customers chosen at random in a supermarket are asked to taste farmhouse cheddar cheese and processed cheddar cheese. Five of the customers say that they prefer the farmhouse cheese. Investigate whether this result leads you to think that, at the 5% significance level, customers of the supermarket prefer farmhouse cheddar cheese to processed cheddar cheese.

2 A coin is tossed 248 times, giving a head on 144 tosses. Investigate at the 5% significance level whether it is reasonable to think that the coin is fair.

3 Records kept in a hospital show that on average 3 out of every 10 patients survive a particular operation. When operating techniques are improved, it is found that, of the next 120 patients, 48 survive the operation. Test (a) at the 2% level, (b) at the 1% level, whether these new techniques have improved the chances of survival.

Miscellaneous Exercise 4

1 A random variable X is distributed $N(20, 9)$. A value of X is found to be 25. Test whether this value is significantly large at the (a) 5% level, (b) 1% level.

2 The national average sick leave for civil servants during the last 12 months was 10·4 days with standard deviation of 2·2 days. A large department finds that in a random sample of 100 civil servants the average sick leave during the last 12 months was 10·9 days. Assuming that the duration of sick leave is normally distributed, test whether this is significantly greater than the national average, stating clearly your null hypothesis, your alternative hypothesis and your significance level.

3 The sex ratio in mink bred in captivity is known to be 6 males to 5 females. A random sample of 6 mink bred in captivity is taken. Find, to three decimal places, the probability that
(a) exactly one will be female,
(b) at least 2 will be female.
To investigate whether the sex ratios are the same for mink bred in captivity and for wild mink, a random sample of 50 wild mink was obtained and it was found that this contained 18 females. Set up an appropriate null hypothesis and an appropriate alternative hypothesis and carry out a 5% significance test, using a normal approximation, to determine whether the sex ratios are equal for the two types of mink.

4 Records of the widths of pins produced on a certain machine indicate that the widths are normally distributed with mean 1·224 mm and standard deviation 0·052 mm. Two hundred samples, each consisting of 100 pins, are chosen at random from the production. Calculate the expected number of the 200 samples having a mean width less than 1·223 mm.
On a certain day it is suspected that the machine is not working properly, the

mean width apparently being changed. A random sample of 100 pins is taken and it is found to have a mean width of 1·244 mm. Assuming that, even though the machine is not working properly, there is no change in the standard deviation, determine a 98% confidence interval for the mean width of pins produced that day.

Hence state whether or not you would conclude, at the 2% significance level, that the machine is not working properly that day.

5 Genetic theory states that the result of a certain cross will produce plants with yellow and with red flowers in the ratio 3 : 1. Twenty of the seeds produced as the result of the cross grow into plants, 17 with yellow flowers and 3 with red. Investigate whether this result supports the theory or whether it suggests that the probability of a red-flowered plant is less than 1/4.

6 In a study on the inheritance of colouring in a certain small mammal, a random sample of 200 specimens was found to contain 146 black and 54 brown individuals. Test, at the 2% level, the null hypothesis that the colours black and brown are inherited in the ratio 2 : 1.

7 The thickness of limestone deposits in an area is known to be normally distributed with mean 6·3 m and standard deviation 0·9 m. Calculate, to three decimal places, the probability of finding a deposit of thickness
(a) greater than 8 m,
(b) between 6 m and 7 m.

In a second area the thickness in metres at 7 positions selected at random were

$$7·9 \quad 6·5 \quad 7·0 \quad 7·4 \quad 7·3 \quad 6·7 \quad 8·1.$$

Investigate, at the 1% significance level, whether there is a significant difference in the mean thickness of limestone deposits in the two areas.

8 The diameters of eggs of gannets are normally distributed with mean 4·88 cm and standard deviation 0·21 cm. Find the probability that an egg chosen at random has a diameter
(a) lying between 4·5 cm and 5 cm,
(b) greater than 4·9 cm.

State the limits within which 95% of the diameters of gannets' eggs may be expected to lie.

A sample of 9 eggs collected from a particular island was found to have diameters, in centimetres, as follows:

$$4·7 \quad 4·6 \quad 5·0 \quad 4·9 \quad 4·6 \quad 4·4 \quad 4·8 \quad 5·0 \quad 5·1.$$

Investigate whether this leads you to think that the diameters of eggs of gannets on this island are smaller than elsewhere.

9 The IQs of children of a certain age group may be assumed to be normally distributed with mean 100 and standard deviation 12. Find the lowest IQ for the top 10% of children in this age group.

Find, to three significant figures, the percentage of children in this age group who would be expected to have an IQ
(a) greater than 109,
(b) lying between 109 and 130.

Calculate the probability that a random sample of 5 children of this age group will contain at least 2 children with IQ greater than 109.

A sample of 5 children in this age group, chosen at random from a particular school, is found to contain 2 children with IQ greater than 109. Investigate whether or not this leads you to think that this sample is biased towards children of high IQ.

10   Spores of a fungus are incubated for 48 hours. Over a long period it is found that the average length of their germ tubes after such incubation is 8·2 with variance 0·052. When a new incubator is used to incubate for 48 hours a sample of 20 spores of the same fungus, they are found to have mean germ tube length of 8·32. Discuss, at the 1% significance level, whether or not the new incubator has caused a significant increase in the mean germ tube length. State carefully H_0, H_1 and, giving your reason, whether you are using a one-tailed or a two-tailed test.

11   The number of bacteria in 1 ml of inoculum is assumed to follow a Poisson distribution with mean 3. Given that at least 2 bacteria are required for a dose of 1 ml to be infective, show that the probability of a dose causing infection is approximately 0·8.

Calculate the probability that, if 7 doses are administered, at least 3 of them will cause infection.

Given that 100 doses are administered and only 72 doses cause infection, discuss the conclusion which you draw about the inoculum.

5 Linear regression

5.1 Bivariate data

In the previous chapters we have considered only distributions of a single variable. Each member of the sample had one, and the same, characteristic recorded. For example, we had samples representing populations of times of travel, of heights, and of many other single variables. We now consider populations and samples where each member has the same two particular characteristics recorded; we call this *bivariate data*, as opposed to the previous *univariate data*. A sample of bivariate data can be illustrated in a sketch using two perpendicular axes, one axis representing one recorded characteristic, and the other axis representing the other characteristic. For a sample of size *n*, we can plot the *n* pairs of recordings on a *scatter diagram*. For example, below are the data from a sample of 10 healthy women between the ages of 20 and 60 years for whom age (in years) and diastolic blood pressure (in mm of mercury) are recorded.

Age	22	28	30	34	36	42	45	48	51	58
Blood pressure	72	74	75	73	78	81	80	84	83	90

If one of the two variables has no error in it (in our case "age") this is the *independent variable* and we can call this variable x. The other variable will

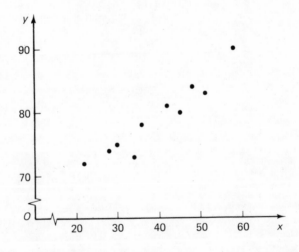

Fig. 5.1

then be y, the *dependent variable* (in our case "blood pressure"), depending on x. The variable y is a random variable, each determination of y having a random error involved in its recording. For our example, the scatter diagram is shown in Fig. 5.1.

In this example, blood pressure can be a function of age (that is, dependent on age), whereas we would never expect age to be dependent on blood pressure. Age is the fixed variable, without error, and we can choose the ages of the women whose blood pressures we intend to measure. Having drawn the scatter diagram we wish to obtain an equation $y = f(x)$ from which we can predict, for a given value of x, age, the expected value of y, blood pressure. Provided that the points in the scatter diagram appear to have roughly a straight-line configuration, we can try to fit a straight-line equation, $y = \alpha + \beta x$, where α and β are to be estimated, $\hat{\alpha}$ and $\hat{\beta}$ being their estimated values.

In practice, many examples do not show a straight-line configuration on the scatter diagram, and then we try to fit some suitable curve (for example, a higher-order polynomial, an exponential or a hyperbolic curve) to the points. The usual method followed is to assume a form of curve which appears to fit the general configuration of the points, and then to transform one or both of the variables to reduce the equation to that of a straight line in the transformed variables. The scatter diagram referred to the transformed variables is then drawn; if the points have a roughly straight-line plot, then the equation of the assumed curve is suitable to use. We then fit a straight-line equation to the transformed data, and finally replace the original variables to obtain $y = f(x)$.

For example, if we feel that the scatter diagram appears to suggest a logarithmic connection between x and y, say $y = \alpha + \beta \ln x$, then we could write $X = \ln x$ and plot a scatter diagram of the points (X, y). If this diagram shows a roughly linear plot, then we fit a straight line $y = \alpha + \beta X$, by the method of least squares which we describe next.

For all A-level examination purposes, the data given will have a scatter diagram with a roughly linear configuration of points, and we fit an approximating straight line directly to the given data. A very rough method of fitting this line would be to draw it by eye using a ruler, but this is obviously not a very satisfactory method since any two people are unlikely to draw the same line. The method used for finding the 'best' fitting straight line, with equation $y = \alpha + \beta x$, is *the method of least squares*. If we imagine this theoretical line, which 'best' fits our sample data points, drawn on to the scatter diagram, as shown in Fig. 5.2, we can then draw a line parallel to the Oy axis from each of the data points on to the theoretical fitted line. Given that the n data points are $(x_1, y_1), (x_2, y_2), \ldots, (x_r, y_r), \ldots, (x_n, y_n)$, then the length, d_r, of the line from (x_r, y_r) to the fitted line is called the *deviation* or *residual* of that point. Some of the lengths $d_r, r = 1, 2, 3, \ldots, n$, will be negative and some positive.

For the method of least squares, we find the estimated values, $\hat{\alpha}$ and $\hat{\beta}$, of α and β so that the sum of the squares of the deviations S, where

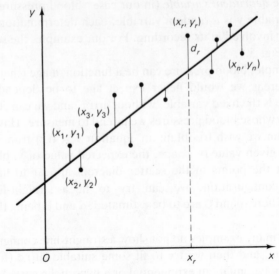

Fig. 5.2

$$S = d_1^2 + d_2^2 + \ldots + d_n^2,$$

is a minimum. We can write down d_r in terms of x_r, y_r, $\hat{\alpha}$, and $\hat{\beta}$, thus

$$d_r = y_r - \hat{\alpha} - \hat{\beta}x_r.$$

Hence,

$$S = \sum_{r=1}^{n} (y_r - \hat{\alpha} - \hat{\beta}x_r)^2.$$

For S to be a minimum, the derivatives of S with respect to $\hat{\alpha}$ and $\hat{\beta}$ must both be zero. The derivative with respect to $\hat{\alpha}$ equal to zero gives

$$0 = -2 \sum_{r=1}^{n} (y_r - \hat{\alpha} - \hat{\beta}x_r),$$

$$\Rightarrow 0 = \sum_{r=1}^{n} y_r - \hat{\alpha} \sum_{r=1}^{n} 1 - \hat{\beta} \sum_{r=1}^{n} x_r,$$

$$\Rightarrow 0 = n\bar{y} - n\hat{\alpha} - \hat{\beta}n\bar{x},$$

since $\quad \bar{y} = \dfrac{\sum\limits_{r=1}^{n} y_r}{n} \quad$ and $\quad \bar{x} = \dfrac{\sum\limits_{r=1}^{n} x_r}{n},$

$$\Rightarrow \bar{y} = \hat{\alpha} + \hat{\beta}\bar{x}.$$

This tells us that the line $y = \hat{\alpha} + \hat{\beta}x$ passes through the point $(\bar{x}, \bar{y})$.

The derivative with respect to β equated to zero gives

$$0 = -2 \sum_{r=1}^{n} x_r(y_r - \hat{\alpha} - \hat{\beta}x_r)$$

$$\Rightarrow 0 = \sum_{r=1}^{n} x_r y_r - \hat{\alpha} \sum_{r=1}^{n} x_r - \hat{\beta} \sum_{r=1}^{n} x_r^2$$

$$\Rightarrow \sum_{r=1}^{n} x_r y_r = \hat{\alpha} n\bar{x} + \hat{\beta} \sum_{r=1}^{n} x_r^2.$$

But, since $\hat{\alpha} = \bar{y} - \hat{\beta}\bar{x}$,

$$\sum_{r=1}^{n} x_r y_r = n\bar{x}\bar{y} - \hat{\beta}(n\bar{x}^2 - \sum_{r=1}^{n} x_r^2).$$

Therefore,

$$\hat{\beta} = \frac{\sum_{r=1}^{n} x_r y_r - n\bar{x}\bar{y}}{\sum_{r=1}^{n} x_r^2 - n\bar{x}^2}.$$

Hence the equation of the 'best fitting' line becomes

$$(y - \bar{y}) = \hat{\beta}(x - \bar{x}),$$

where $\hat{\beta}$ is given above. This equation is called *the regression equation of y on x*, and its slope, $\hat{\beta}$, is called the *regression coefficient*. The form in which we obtained $\hat{\beta}$, that is,

$$\hat{\beta} = \frac{\sum_{r=1}^{n} x_r y_r - n\bar{x}\bar{y}}{\sum_{r=1}^{n} x_r^2 - n\bar{x}^2},$$

is probably as good as any for the purpose of calculating $\hat{\beta}$ for a given sample of n points. However, we can convert it to another form which we use when discussing correlation in Chapter 6.

In Chapter 1, we showed that

$$\sum_{r=1}^{n} (x_r - \bar{x})^2 = \sum_{r=1}^{n} x_r^2 - n\bar{x}^2 = ns_x^2,$$

where s_x^2 is the variance of the x readings.

Similarly, it can be shown that

$$\sum_{r=1}^{n} (x_r - \bar{x})(y_r - \bar{y}) = \sum_{r=1}^{n} x_r y_r - n\bar{x}\bar{y}.$$

Substituting these results into the formula for $\hat{\beta}$ we have

$$\hat{\beta} = \frac{\sum_{r=1}^{n} (x_r - \bar{x})(y_r - \bar{y})}{\sum_{r=1}^{n} (x_r - \bar{x})^2}.$$

The expression $\dfrac{\sum\limits_{r=1}^{n}(x_r - \bar{x})(y_r - \bar{y})}{n}$ is called *the covariance of x and y*, and

we give it the symbol s_{xy}. You will notice that this covariance may be either positive or negative, whereas variance is always positive. Then $\hat{\beta} = s_{xy}/s_x^2$, where s_x^2 denotes the variance of the x's.

When these values for $\hat{\beta}$ and for $\hat{\alpha} = \bar{y} - \hat{\beta}\bar{x}$ are substituted into S, the sum of squares of the deviations, we obtain the minimum value of S. This is *the minimum sum of squares of the residuals* and is given by

$$\sum_{r=1}^{n}(y_r - \bar{y})^2 - \frac{\left[\sum\limits_{r=1}^{n}(x_r - \bar{x})(y_r - \bar{y})\right]^2}{\sum\limits_{r=1}^{n}(x_r - \bar{x})^2} = ns_y^2 - \frac{n^2(s_{xy})^2}{ns_x^2}$$

$$= n\left[s_y^2 - \frac{(s_{xy})^2}{s_x^2}\right],$$

where s_x^2 is the variance of the x readings,
s_y^2 is the variance of the y readings,
s_{xy} is the covariance of x and y.

The more closely grouped the data points are about the regression line, the smaller will be the minimum sum of the squares of the residuals, and the more accurate will be the fit of the regression line.

We have found estimates $\hat{\alpha}$ and $\hat{\beta}$ from our sample points and these can be shown to be unbiased estimates of α and β, although this is beyond the scope of this text. Later, in Chapter 7, we will discuss how well this fitted line represents the true line for the population from which the sample is taken. Assuming that the equation $y = \hat{\alpha} + \hat{\beta}x$ represents the true line for the population, we can make an estimate of y for any given value of x, *but not for x given y since y is dependent on x.*

It is, however, very unwise to attempt to predict y for values of x which are not within, or very close to, the range of the sample data, for we have little or no knowledge as to whether conditions are the same within and outside that range. Outside the range, conditions could be materially different. For example, in the age and blood pressure example, it would be unacceptable to make a prediction of blood pressure for a woman of 90 years from the regression line of the sample data without knowing whether or not there is a levelling off or alternatively a large increase in blood pressure in old age. We could however expect to predict from our equation the blood pressure of a healthy woman aged 38 years, or even of one aged 62 years.

There are two main methods of performing the calculations for $\bar{x}$, $\bar{y}$ and $\hat{\beta}$, which are the quantities we must calculate in order to write down the equation of the regression line. The following example is worked using both methods.

Example 1 The length, in centimetres, of a piece of fine straight wire was recorded for various temperatures, measured in degrees centigrade (°C). The results are shown below. Find the equation of the regression line of length on temperature.

Using this equation, estimate, to two decimal places, the length of the wire when the temperature is 103·3°C.

Temperature	103·0	103·5	104·0	104·5	105·0
Length	8·05	8·10	8·17	8·25	8·33

For either method, we start by drawing a scatter diagram, Fig. 5.3. The independent variable, x, is the temperature; the dependent variable, y, is the length.

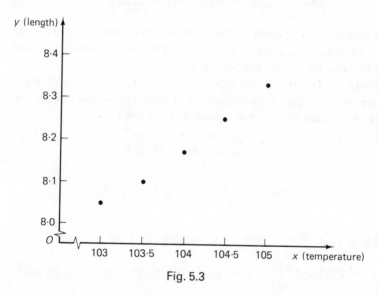

Fig. 5.3

The data points have an approximately linear configuration and so it is reasonable to fit a straight-line regression.

Method 1 We set up a table of values in order to organise the summations required to calculate $\bar{x}$, $\bar{y}$ and $\hat{\beta}$, although these can be found on some calculators without the need to set up a table.

x	y	$(x - \bar{x})$	$(x - \bar{x})^2$	$(y - \bar{y})$	$(x - \bar{x})(y - \bar{y})$
103·0	8·05	−1	1	−0·13	0·13
103·5	8·10	−0·5	0·25	−0·08	0·04
104·0	8·17	0	0	−0·01	0
104·5	8·25	0·5	0·25	0·07	0·035
105·0	8·33	1	1	0·15	0·15
520	40·9		2·50		0·355

$$\bar{x} = \frac{520}{5} = 104, \qquad \bar{y} = \frac{40\cdot9}{5} = 8\cdot18, \qquad \hat{\beta} = \frac{0\cdot355}{2\cdot50} = 0\cdot142.$$

Therefore, the regression equation of length on temperature is given by

$$(y - 8\cdot18) = 0\cdot142(x - 104),$$

that is,
$$y = 0\cdot142x - 6\cdot588.$$

When $x = 103\cdot3$, $y = 8\cdot0806 \approx 8\cdot08$ cm.

For this method, the two summations $\sum_{r=1}^{n} (x_r - \bar{x})^2$ and $\sum_{r=1}^{n} (x_r - \bar{x})(y_r - \bar{y})$ can be evaluated directly, as shown above, or we could have used the results

$$\sum_{r=1}^{n} (x_r - \bar{x})^2 = \sum_{r=1}^{n} x_r^2 - n\bar{x}^2,$$

and
$$\sum_{r=1}^{n} (x_r - \bar{x})(y_r - \bar{y}) = \sum_{r=1}^{n} x_r y_r - n\bar{x}\bar{y},$$

if this makes the arithmetic simpler. However, if large numbers are involved in x or y values, care must be taken so that the use of these latter forms does not introduce large rounding-off errors.

Method 2 The regression equation is $y = \hat{\alpha} + \hat{\beta}x$. Multiplying by x we have $xy = \hat{\alpha}x + \hat{\beta}x^2$. Substituting the n data values and adding for all the points, we obtain the *normal equations* for y on x:

$$\sum_{r=1}^{n} y_r = n\hat{\alpha} + \hat{\beta} \sum_{r=1}^{n} x_r,$$

and
$$\sum_{r=1}^{n} x_r y_r = \hat{\alpha} \sum_{r=1}^{n} x_r + \hat{\beta} \sum_{r=1}^{n} x_r^2.$$

We know that $\sum_{r=1}^{n} x_r = 520$, $\sum_{r=1}^{n} y_r = 40\cdot9$, and we can calculate

$\sum_{r=1}^{n} x_r y_r = 4253\cdot955$, $\sum_{r=1}^{n} x_r^2 = 54\,082\cdot5$. Substituting these values into the two normal equations we have

$$40\cdot9 = 5\hat{\alpha} + 520\hat{\beta},$$
$$4253\cdot955 = 520\hat{\alpha} + 54\,082\cdot5\hat{\beta}.$$

Eliminating $\hat{\alpha}$ gives

$$4253\cdot955 = 104(40\cdot9 - 520\hat{\beta}) + 54\,082\cdot5\hat{\beta}$$
$$\Rightarrow 4253\cdot955 - 4253\cdot6 = 2\cdot5\hat{\beta}$$
$$\Rightarrow \hat{\beta} = \frac{0\cdot355}{2\cdot5} = 0\cdot142$$

as before.

The regression equation is

$$(y - 8\cdot18) = 0\cdot142(x - 104),$$

or,
$$y = 0\cdot142x - 6\cdot588.$$

Exercise 5.1

1 Given the following 12 pairs of values of x and y

x	84	82	82	85	89	90	88	92	83	89	98	99
y	78	77	85	88	87	82	81	77	76	83	97	93

draw a scatter diagram for the data.
  Find the equation of the regression line of y on x.
  From your equation estimate y when $x = 87$.

2 A market gardener growing tomato plants in a greenhouse divided up the plants into 6 equal groups. He treated each group with a different concentration of fertiliser. The average crop from each group of plants is shown below.

Concentration C (gl^{-1})	$\frac{1}{2}$	1	2	3	4	6
Crop R (kg)	5	8	13	18	25	36

Draw a scatter diagram for these data, and mark on your diagram the point representing the mean of the data.
  Find the equation of a suitable regression line from which the crop to be expected for a concentration of $2 \cdot 5$ gl^{-1} can be estimated, and give the value of this expected crop. Sketch the regression line on your scatter diagram.
  Calculate the sum of squares of the residuals and explain what this value represents with regard to your regression line.

Miscellaneous Exercise 5

1 From ten pairs of values of the variables x and y, find the line of regression of y on x given that

$$\Sigma x = 35, \quad \Sigma y = 95, \quad \Sigma x^2 = 250, \quad \Sigma xy = 475.$$

2 The following data are the rates of oxygen consumption of a particular mammal at different room temperatures:

Temperature (°C)	-18	-15	-10	-5	0	5	10	19
Oxygen consumption (ml g^{-1} h^{-1})	5·2	4·7	4·5	3·6	3·4	3·1	2·7	1·8

State which is the dependent and which is the independent variable.
  Draw a scatter diagram.
  Find the equation of the line of regression of oxygen consumption rate on temperature and estimate the oxygen consumption rate at 2°C.

3 Thin sections of a plant stem are immersed in a salt solution for varying lengths of time (in hours). The salt content (milligrams per 1000 g of water) in the stem tissues after immersion is shown below. Find the equation of the line of regression of salt content on time.
  From your equation deduce the average amount of increase of salt content per hour.

Time (hours)	21·0	46·0	67·0	90·0	96·0
Salt content (mg/1000 g)	7·2	11·0	14·0	19·1	20·0

4 In the following table M grams is the mass of a chemical dissolved in water at $T°C$:

T	10	20	30	40	50	60	70	80
M	44	47	50	55	60	63	64	65

Draw a scatter diagram for these data.

Show that the covariance is equal to 172·5.

Find the equation of the line of regression of M on T.

5 The decolourisation of a starch–iodine complex varies with the amount of ascorbic acid present in a given volume of solution. Readings of decolourisation, D, for given amounts, A, of acid are shown below. Show that the covariance of D and A is −308·6. Find the equation of the regression line of D on A.

A (μg/ml)	100	200	300	400	500	600	700
D		6·0	5·5	4·8	4·0	3·2	2·4	1·4

6 A weight W (newtons) is attached to a spring which hangs vertically. The extension e (cm) is shown in the following table for varying values of W. Show that the covariance of W and e is 8·98.

Find the equation of the line of regression of e on W. Estimate the average increase in the extension per unit increase in the attached weight.

W	1	2	3	4	5	6	7	8
e		1·9	3·8	4·5	6·9	8·6	11·1	12·0	13·5

7 The variables V and T are known to be linearly related. Sixty pairs of experimental observations of the two variables give the following results:

$$\Sigma V = 83\cdot 4, \quad \Sigma T = 402\cdot 0, \quad \Sigma V^2 = 384\cdot 6, \quad \Sigma T^2 = 2700\cdot 4, \quad \Sigma VT = 680\cdot 20.$$

Obtain the equation of the regression line from which we can estimate V when T has the value 5·8, and give the value of this estimate.

8 The body and liver masses of ten 8-month-old mice are shown below:

Body mass (x grams)	22	25	32	33	27	31	27	28	33	39
Liver mass (y milligrams)	98	116	136	130	120	135	137	94	124	150

Draw a scatter diagram.

Calculate the mean values of the variables and show the point representing this mean on your diagram.

Calculate the equation of the regression line of y on x, and draw this line on your scatter diagram.

9 In an experiment the values of a variable p (density) were plotted against the natural logarithm of the tensile stress T. The following results were obtained:

p ($g\,cm^{-3}$)	80	90	100	110	120	130	140	150
$\ln T$ ($g\,cm^{-2}$)	2·6	1·4	4·5	4·9	6·5	5·2	7·0	7·5

Calculate the line of regression of $\ln T$ on p, and express T in terms of p.

Hence estimate the value of T when $p = 125$.

10 On six consecutive days the oxygen concentration C ($mg\,l^{-1}$) in water from a well was recorded. The values for $\ln(10 - C)$ are shown below. During those six days oxygen was being pumped into the well water.

t (number of days during which oxygen is being pumped in)	1	2	3	4	5	6
$\ln (10 - C)$	2·104	1·946	1·792	1·629	1·526	1·335

Assuming that $10 - C = me^{-kt}$, where m is a constant, find the line of regression of $\ln (10 - C)$ on t.

Hence estimate the values of k and m.

Estimate, also, the value of C when $t = 7$, assuming that the pumping of oxygen into the well continues after the six days.

6 Correlation

6.1 The product moment correlation coefficient

In Chapter 5 we saw that, when a variable y is dependent on a non-random variable x, we can find the equation of the regression line of y on x, using the method of least squares, and, from this equation, estimate the value of y for a given x value. If we have a sample of data giving us the leg and arm lengths of 20 men, we cannot say that leg length is dependent on arm length, nor that arm length is dependent on leg length. In a problem of this kind, all we can consider is the amount of relationship between the two variables, arm length and leg length. This is a problem of *correlation*; we try to answer the question 'Is there any relationship between the two variables and, if so, to what degree are they related?' To do this, we try to determine how well an equation (and we consider only linear equations) represents the relationship between the two variables.

As in a regression problem, we can start by drawing a scatter diagram for the n points (x_r, y_r), but now it does not matter which axis is used to represent either variable, provided that each axis is clearly marked to show which of the variables it represents. When we look at the scatter diagram, if y tends to increase with x we have *positive correlation*, if y tends to decrease as x increases we have *negative correlation*, and if there is just a shapeless scatter of points we have *no correlation*. We concern ourselves only with cases of correlation in which a *linear* relationship is indicated between the two variables.

Let the two variables be u and v. If we consider u as being dependent on v, then we could find, using the method of least squares described in Chapter 5, the regression equation of u on v in the form

$$(u - \bar{u}) = \hat{\beta}_1(v - \bar{v}) = \frac{s_{uv}}{s_v^2}(v - \bar{v}).$$

Similarly, if we consider v as being dependent on u, then we could find the regression equation of v on u in the form

$$(v - \bar{v}) = \hat{\beta}_2(u - \bar{u}) = \frac{s_{uv}}{s_u^2}(u - \bar{u}).$$

In general, these are two distinct equations of two different straight lines both passing through the point $(\bar{u}, \bar{v})$, as shown in Fig. 6.1.

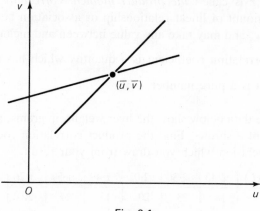

Fig. 6.1

We can write the two equations in a slightly different form by dividing the former throughout by s_u, the standard deviation of the u values, and the latter by s_v, the standard deviation of the v values. We then have

$$\frac{u - \bar{u}}{s_u} = \frac{s_{uv}}{s_u s_v}\left(\frac{v - \bar{v}}{s_v}\right),$$

$$\frac{v - \bar{v}}{s_v} = \frac{s_{uv}}{s_u s_v}\left(\frac{u - \bar{u}}{s_u}\right).$$

You will notice that we have normalised the variables in the same way that we transform the variable from X to Z by the transformation $Z = \dfrac{X - \mu}{\sigma}$ when dealing with a normal distribution. We will use U and V for these normalised variables; that is, we write $U = \dfrac{u - \bar{u}}{s_u}$, $V = \dfrac{v - \bar{v}}{s_v}$. The equations of the two lines can then be written as

$$U = rV, \qquad V = rU,$$

where

$$r = \frac{s_{uv}}{s_u s_v} = \frac{\text{covariance}}{(\text{SD of } u \text{ values})(\text{SD of } v \text{ values})}.$$

When $r = 1$, the equations both become $U = V$, the two lines in Fig. 6.1 coincide, and we have *perfect positive correlation*.

When $r = 0$, the equations become $U = 0$ and $V = 0$, the two lines in Fig. 6.1 are now at right angles with one parallel to the u axis and the other parallel to the v axis, and we have *no correlation*.

When $r = -1$, the equations both become $U = -V$, the two lines in Fig. 6.1 coincide but both have a negative slope, and we have *perfect negative correlation*.

The quantity r is called *the product moment correlation coefficient*. It measures the amount of linear relationship or association between the two variables u and v, and may take any value between and including ± 1. Since $r = \dfrac{S_{uv}}{S_u S_v}$, the correlation coefficient is a quantity which has no dimensions (units), that is it is a pure number.

Example 1 The data below show the liver weight, in grams, and the length, in centimetres, of 8 shrews. Find the product correlation coefficient.
State any conclusion which you draw from your result.

Liver weight (L)	2·40	2·30	2·02	1·40	1·59	1·26	2·41	2·30
Length (B)	18	23	20	13	24	20	22	20

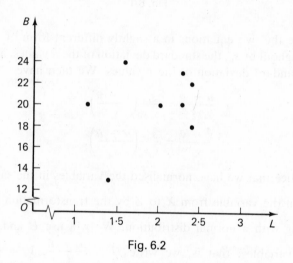

Fig. 6.2

The scatter diagram, Fig. 6.2, in which we could equally well have represented length and liver weight along the 'horizontal' and 'vertical' axes respectively, does not appear to show much, if any, correlation between the variables. However, if we calculate r, we can then make a more definite judgement about the presence or absence of linear correlation.

L	B	$(L - \bar{L})$	$(L - \bar{L})^2$	$(B - \bar{B})$	$(B - \bar{B})^2$	$(L - \bar{L})(B - \bar{B})$
2·40	18	0·44	0·1936	−2	4	−0·88
2·30	23	0·34	0·1156	3	9	1·02
2·02	20	0·06	0·0036	0	0	0
1·40	13	−0·56	0·3136	−7	49	3·92
1·59	24	−0·37	0·1369	4	16	−1·48
1·26	20	−0·70	0·4900	0	0	0
2·41	22	0·45	0·2025	2	4	0·90
2·30	20	0·34	0·1156	0	0	0
15·68	160		1·5714		82	3·48

$$\bar{L} = 1 \cdot 96, \ \bar{B} = 20,$$

$$r = \frac{3 \cdot 48}{\sqrt{82} \times \sqrt{1 \cdot 5714}} = 0 \cdot 307.$$

This a low value of r and hence we conclude that there is little or no *linear* relationship between the two variables.

When we obtain a value of r which is close to zero we can say, with some certainty, that there appears to be little or no correlation between the variables. Similarly, if the value of r is close to $+1$ or to -1 we can say that there appears to be good correlation, positive or negative respectively. Suppose, however, the result for r had been $r \approx 0 \cdot 6$. Then we can say that there appears to be some degree of correlation but we cannot draw any definite conclusion. In such cases, *provided that the variables are normally distributed*, we can perform a significance test in which we test the null hypothesis $r = 0$. That is, we test the hypothesis that, for the population, there is no correlation. This significance test requires the use of a *t*-test, which is discussed in Chapter 7.

There are some points which must always be borne in mind when calculating and interpreting a correlation coefficient.

(i) The value of r, the product moment correlation coefficient, measures only the degree of *linear* relationship between the two variables. A value of r near to or equal to zero does not mean that the two variables are not related, but that they are not linearly related and there might be, for example, a logarithmic or a cubic relationship between the two.

(ii) Although we may find a value of r near to $+1$ or to -1 and hence a good linear relationship between the two variables, this does *not* mean that either is caused by the other. There could well be a third variable, or even several variables, causing the two variables to change in the same, or in an inverse, way. For example, if we plotted the number of reported burglaries against the amount of wine drunk over the past 10 years in England, it might very well show a high degree of correlation, but in no way could we say that the wine consumption is causing the burglaries nor vice versa. There may be outside social factors which are affecting both sets of statistics in the same way.

Exercise 6.1

1 The total rainfall in millimetres at Lerwick and at Gatwick for 12 successive months was

Lerwick	125	58	70	27	39	32	100	57	46	88	44	110
Gatwick	96	103	70	50	55	46	24	76	52	60	35	87

Find the product moment correlation coefficient.

2 Find the covariance and the product moment correlation coefficient for the following 10 pairs of numbers x and y:

x	102	102	105	105	107	107	101	106	96	81
y	95	97	99	102	94	97	94	97	92	90

3 The table below shows the length x mm of the wings and the length y mm of the antennae of 9 butterflies of a certain species:

x	10·8	10·8	11·2	10·3	10·2	10·7	10·8	10·2	10·4
y	7·8	7·9	7·7	7·4	7·1	7·4	7·6	7·2	7·4

Given that $\Sigma x^2 = 1012·18$, $\Sigma y^2 = 506·83$, $\Sigma xy = 716·10$, calculate the product moment correlation coefficient for these data.

6.2 Rank correlation

Ranked data

All the data we have met so far have been from samples in which the variables were measured or counted so that the data consisted of numerical readings or recordings. Sometimes, for example in a beauty contest or in some scholarship examinations, the members of the sample are given only *ranks* (rather than numerical values). We say that someone is first, rank 1, the next is second, rank 2, and so on until the last, nth, person is rank n. That is, we know only the relative positions of the members of the sample. On other occasions, although the sample data consist of numerical readings, we may decide that we prefer to rank the data and to perform statistical calculations with the rankings rather than with the original data. Of course, this tends to make the work less accurate. For example, if the original data are 23, 2 and 26, these have ranks 2, 1 and 3 respectively, but we would still have the same rankings 2, 1 and 3 had the middle reading been, say, 22. However, provided this loss of accuracy is acceptable, calculations with ranks usually involve simpler arithmetic than those with the original data.

When ranking a set of sample data, we may find that we have two or more equal readings. When this happens, we give each of them the mean of the ranks which they would occupy. For example, ranking the sample of nine values

$$0·2 \quad 0·8 \quad 0·6 \quad 0·5 \quad 0·8 \quad 0·8 \quad 1·2 \quad 0·3 \quad 0·5$$

we have ranks

$$1 \quad 7 \quad 5 \quad 3\tfrac{1}{2} \quad 7 \quad 7 \quad 9 \quad 2 \quad 3\tfrac{1}{2}$$

since the two values of $0·5$ would fill the ranks 3 and 4 and so both take the rank $3\tfrac{1}{2}$. Similarly the three values of $0·8$ would fill the ranks 6, 7 and 8 and so each takes the rank 7.

Another occasion on which we might decide to use ranks rather than the original data is when the statistical test which we wish to perform requires that the population from which the sample comes has some given form of distribution. (Often, the requirement is that the distribution is normal.) In §6.1 we said that we can perform a significance test on r when its value is not close to 0 or to ± 1 using a t-test (Chapter 7). This test requires that the population is normally distributed, and if we cannot, or do not wish to, make this assumption then we must rank the data and calculate a *rank correlation coefficient* instead of a product moment coefficient. The corresponding significance tests for rank correlation coefficients require no such assumption. We describe here the methods by which two rank correlation coefficients can be obtained and tested for significance. Their derivation is beyond the scope of this book.

Spearman's rank correlation coefficient, r_S.

The formula for r_S for a sample of n pairs of ranks is derived by obtaining a product moment correlation coefficient on the two sets of rankings 1 to n for the two variables. We do not show this derivation but quote the result:

$$r_S = 1 - \frac{6\sum_{r=1}^{n} d_r^2}{n(n^2 - 1)},$$

where d_r is the difference between the ranks of the rth pair of readings.

Example 2 In a test, 9 students were given marks both for practical work and for theoretical answers. The results are shown below. Obtain
(a) the product moment correlation coefficient,
(b) Spearman's rank correlation coefficient.
   Discuss your result in each case.

Practical (P)	11	14	14	10	7	12	9	13	18
Theoretical (T)	12	14	13	7	8	11	10	15	19

(a)
$$\sum_{r=1}^{9} P_r = 108 \Rightarrow \bar{P} = 12.$$

$$\sum_{r=1}^{9} T_r = 109 \Rightarrow \bar{T} = \frac{109}{9}.$$

$$\sum_{r=1}^{9} (P_r - \bar{P})^2 = 1 + 4 + 4 + 4 + 25 + 0 + 9 + 1 + 36 = 84.$$

$$\sum_{r=1}^{9} (T_r - \bar{T})^2 = \sum_{r=1}^{9} T_r^2 - 9\bar{T}^2$$

$$= [144 + 196 + 169 + 49 + 64$$

$$+ 121 + 100 + 225 + 361] - 9\left(\frac{109}{9}\right)^2$$

$$= 1429 - \frac{11\,881}{9} = \frac{980}{9}.$$

$$\sum_{r=1}^{9}(T_r - \bar{T})(P_r - \bar{P}) = \sum_{r=1}^{9}T_rP_r - 9\bar{T}\bar{P} = 87.$$

Therefore $$r = \frac{87 \times 3}{\sqrt{84} \times \sqrt{980}} = 0\cdot910.$$

This value is close to $+1$, so we can say that a strong positive linear relationship exists between abilities in practical and in theoretical work, the abilities in the two types of work increasing together.

(b) Putting the data into ranks from the highest mark, rank 1, to the lowest mark, rank 9, we obtain the following table of ranks.

P	6	$2\frac{1}{2}$	$2\frac{1}{2}$	7	9	5	8	4	1
T	5	3	4	9	8	6	7	2	1
d_r	1	$-\frac{1}{2}$	$-1\frac{1}{2}$	-2	1	-1	1	2	0

$$\sum_{r=1}^{9}d_r^2 = 1 + \tfrac{1}{4} + \tfrac{9}{4} + 4 + 1 + 1 + 1 + 4 + 0 = 14\cdot5$$

$$\Rightarrow r_S = 1 - \frac{6 \times 14\cdot5}{9 \times 80} = 0\cdot879.$$

Again, we have found a value of r_S close to $+1$, thereby showing a strong positive correlation between the rankings in the two types of work. The ranks of P and of T increase together in a linear fashion.

For a Spearman correlation coefficient, if we are in doubt about our decision, we can look up the significance of the result for r_S in tables. We perform a significance test, testing the null hypothesis $H_0 : r_S = 0$ (that is, no correlation) against $H_1 : r_S > 0$. The tables available may be either of critical values for r_S, or of probabilities associated with $\sum_{r=1}^{9} d_r^2$. We now indicate how we use both types of table.

For Example 2, we had $r_S = 0\cdot879$, $\sum_{r=1}^{9} d_r^2 = 14\cdot5$, and $n = 9$. In a table of critical values for r_S we find that the 5% level is $r_S = 0\cdot600$ and the 1% level is $r_S = 0\cdot783$. Hence our value is greater than the 1% critical level for r_S and so we reject H_0 strongly. In a table of probabilities associated with $\sum_{r=1}^{9} d_r^2$ in Spearman's coefficient, we find that for $n = 9$ and $\sum_{r=1}^{9} d_r^2 = 14\cdot5$, the probability is less than $0\cdot0023$, which is very small, and again we reject strongly the null hypothesis of no correlation between the two sets of ranks.

Kendall's rank correlation coefficient, r_K

We start by placing one of the samples in rank order, $1, 2, \ldots, n$. Underneath each rank we write down the rank of the corresponding member of the other sample.

Starting with the left-hand member, say k_1, of the second row, we write down $+1$ for every rank on the right of k_1 which is greater than k_1 and -1 for every rank on the right of k_1 which is less than k_1. We then move on to the next entry (k_2) in the second row and repeat the process with all the ranks to the right of k_2. We carry on in this way for all the entries in the second row and then add together all the values of $+1$ and -1 to obtain a total, S. Kendall's rank coefficient, r_K, is *defined* by

$$r_K = \frac{2S}{n(n-1)},$$

where n is the sample size.

Example 3 For the data of Example 2, find r_K and interpret your result.

The ranks were

P	6	$2\frac{1}{2}$	$2\frac{1}{2}$	7	9	5	8	4	1
T	5	3	4	9	8	6	7	2	1

Putting the T ranks in order and writing down the corresponding P ranks beneath them, we have

T	1	2	3	4	5	6	7	8	9
P	1	4	$2\frac{1}{2}$	$2\frac{1}{2}$	6	5	8	9	7

For the second, P, row we have

1: $\ +1+1+1+1+1+1+1+1 = 8$,   since each of the ranks to the right of the first entry is greater than 1.

4: $\ -1-1+1+1+1+1+1 = 3$,   since the first two to the right are less than 4, the rest are greater than 4.

$2\frac{1}{2}$: $\ 0+1+1+1+1+1 = 5$,

$2\frac{1}{2}$: $\ 1+1+1+1+1 = 5$,

6: $\ -1+1+1+1 = 2$,

5: $\ 1+1+1 = 3$,

8: $\ 1-1 = 0$,

9: $\ -1 = -1$.

Total, $S = 25$.

Hence

$$r_K = \frac{2 \times 25}{9 \times 8} = 0 \cdot 694.$$

We then look up the significance of this result in tables. The null hypothesis that we are testing is $H_0 : r_K = 0$, against $H_1 : r_K > 0$. Again, the tables may be of two kinds. The first kind is a table of critical values for r_K, and in this we find that for $n = 9$, $r_K = 0.694$, the 1% critical level is $r_K = 0.6667$. Hence our value of r_K is greater than the 1% critical level and we reject H_0 strongly. The other kind of table shows the probabilities associated with S, and, in this we find that for $n = 9$, $S = 25$, the probability is less than 0.0063, showing a highly significant result. There appears to be strong positive correlation, showing a strong linear relationship between the two sets of ranks, the ranks for practical and for theoretical work increasing together.

Exercise 6.2

1 Nine students take a test on entering a car maintenance course and they are ranked 1 to 9 from best to worst. At the end of the course they take another test and are ranked in the same way. The rankings are shown below.

Student	A	B	C	D	E	F	G	H	J
Start of course	1	2	3	4	5	6	7	8	9
End of course	5	3	1	8	2	6	4	9	7

Find, to two decimal places, (a) Spearman's coefficient, (b) Kendall's coefficient of rank correlation.

   Referring to a table of critical values, state briefly whether or not you conclude from your result that the relative achievements of the students at the beginning and end of the course are correlated.

2 A girl was asked to place in rank order of age 8 dolls, giving rank 1 to the doll which she thought was the oldest. The actual dates of the dolls, in the order in which the girl put them, were

$$1837, \quad 1848, \quad 1821, \quad 1840, \quad 1898, \quad 1880, \quad 1935, \quad 1923.$$

Calculate a coefficient of correlation, and comment on the value which you obtain.

3 Two television viewers were asked to give marks out of twenty to ten TV programmes. The marks they awarded are shown below. Calculate a rank correlation coefficient for the data and comment on the agreement of the rankings.

Programme	A	B	C	D	E	F	G	H	J	K
Viewer 1	17	9	8	20	19	15	10	18	9	16
Viewer 2	13	10	12	19	17	18	14	15	7	9

Miscellaneous Exercise 6

1 The data shown below are the heights of 10 mothers and their oldest adult daughters.

Height of mother (inches)	64	66	64	61	64	62	63	61	67	65
Height of daughter (inches)	65	67	66	62	61	59	63	60	64	64

Draw a scatter diagram of the pairs of readings. Calculate the product moment correlation coefficient.

Express the data in ranks and find a coefficient of rank correlation.

Referring to a table of critical values, state any conclusion which you draw from your result.

2   In each of the months October to March inclusive, the number, N, of road accidents in a town and the number, W, of wet days in that month are recorded. From the data in the table, obtain the equation of the line of regression of N on W.

Calculate the product moment correlation coefficient between N and W, and comment on the value which you obtain.

W	12	20	17	10	13	11
N	20	48	44	15	22	18

3   Calculate the product moment correlation coefficient for the following data.

Year	1977	1978	1979	1980	1981	1982
Turnover (£ million)	34·5	37·7	35·9	39·7	40·8	41·2
Profit (£ million)	4·6	4·4	4·2	4·9	5·4	4·7

Calculate also Spearman's rank correlation coefficient and, using a table of critical values, comment on your result, using a 5% significance level.

4   Given that r denotes the product moment correlation coefficient between two discrete variables x and y, state the conclusion that may be drawn if $r = 1$.

Calculate a rank correlation coefficient for the following data which show the number of fatal road accidents, $N \times 10^2$, and the number of licensed vehicles, $M \times 10^5$, in six different countries in 1980. Comment on your result.

Country	A	B	C	D	E	F
Fatal road accidents (N)	11	83	70	19	12	3
Vehicles (M)	16	120	88	20	15	5

5   Eight corresponding pairs of two variables X and Y are given in the following table:

X	3	19	20	25	28	11	12	10
Y	4	18	29	37	31	27	2	4

Draw a scatter diagram for these data.

Find the equations of the regression lines of Y on X and of X on Y and draw these lines on your diagram.

Calculate the product moment correlation coefficient between X and Y.

6   Rock samples were analysed and contained two sulphides x and y in the percentages shown below.

x	0·80	0·51	0·65	0·70	0·40	1·03	0·90	0·61
y	1·05	0·70	0·81	0·99	0·72	0·90	1·15	0·88

Find the product moment correlation coefficient.

Express the data in ranked form and determine a coefficient of rank correlation, and, using tables, discuss the significance of your result.

State the circumstances under which you would use a rank correlation coefficient rather than a product moment correlation coefficient.

7   The total harvest of turnips in England was measured over 5 consecutive years, which were ranked according to the total number of hours of sunshine in each of

the years. The results are shown below. Calculate, to two decimal places, a rank correlation coefficient for these data.

Discuss the significance of your result using a table of critical values relevant to your rank coefficient.

Rank of amount of sunshine	1	2	3	4	5
Crop (10^3 tonnes)	52	63	73	62	77

8　It is believed that a patient who reacts badly to a particular drug on one occasion will do so on the next occasion. Tests on 10 patients gave the results shown below for the number of days that the reaction was apparent. Calculate Spearman's rank correlation coefficient, and use a table of critical values to decide whether the belief appears to be justified.

Patient	1	2	3	4	5	6	7	8	9	10
First occasion	3	10	13	4	4	7	10	9	7	6
Second occasion	2	7	12	1	6	5	8	9	8	9

9　The death rates R_1 and R_2 (deaths per year per thousand of the population) in a certain district from pneumoconiosis and from lung cancer respectively, grouped by age, are as follows:

Age group	20–24	25–29	30–34	35–39	40–44
R_1	3·7	3·2	6·1	8·4	12·1
R_2	2·4	2·8	4·6	4·6	5·4

Age group	45–49	50–54	55–59	60–64	65+
R_1	11·3	16·4	18·9	18·3	19·8
R_2	6·4	6·2	7·9	9·0	8·8

Calculate a rank correlation coefficient between R_1 and R_2.

Using a table of critical values state the conclusion you draw from your result.

10　On the basis of the data shown below, a psychologist concludes that, the older a teenager becomes, the fewer mistakes she makes when sight reading a given song which she has not previously heard.

Age (years)	13	13	14	14	15	16	16	17	18	19
Number of mistakes	9	7	6	7	8	7	4	5	6	3

Draw a scatter diagram of the data.

Find the product moment correlation coefficient and, without further calculation, state whether you conclude that the psychologist's conclusion is valid.

Find the equation of the regression line of mistakes upon age, and predict the expected number of mistakes for a sight reader aged 15 years.

7 Further tests of significance

7.1 Introduction

The significance tests described in this chapter are useful tools for workers in many fields but the mathematics involved in the derivation of these tests is too difficult for any mathematical approach to be attempted in this book. Here each test is presented as a formula with the appropriate conditions under which the test is applicable to a problem. Also, each test is illustrated by worked examples. It is not possible, at this stage, to investigate the distributions upon which the tests are based.

7.2 'Student's' *t*-test, a significance test for a mean

In Chapter 4 we discussed the significance testing of a population mean using large samples. We stated that, if the population is normally distributed, $N(\mu, \sigma^2)$, then the distribution of means of samples of size n is distributed $N(\mu, \sigma^2/n)$ and hence, for all values of n, we can perform a significance test on μ using normal area tables *provided σ is known*. Further we stated that, *if n is large*, the distribution of sample means is approximately normally distributed $N(\mu, \sigma^2/n)$ and we can perform a significance test using normal area tables even if σ is not known and we have to use the estimated value $\hat{\sigma}$.

For the case of a *small sample of size n from a normal distribution for which we do not know σ* the use of normal area tables is too inaccurate. It is for such cases that we use Student's *t*-test, to test the hypothesis that a normal population has mean μ.

In 1908 the statistician W. S. Gosset (who published under the pseudonym 'Student') found the equation of the distribution of the statistic t given by

$$t = \frac{\bar{x} - \mu}{\hat{\sigma}/\sqrt{n}},$$

where $\bar{x}$ is the mean of a sample of size n and $\hat{\sigma}^2$ is the unbiased estimate of the population variance. The distribution of t is symmetrical, similar to the shape of a normal curve but 'flatter', is dependent on the value of n, and, as n increases, tends to the normal distribution $N(0, 1)$. Figure 7.1 shows some distributions of t for increasing n values.

Tables of areas (probabilities) under the t curve are published; these tables, unlike those for the curve $N(0, 1)$ in normal area tables, have to take into account the changing t curve for different values of n. The tables show the probabilities, the significance levels and v, *the degrees of freedom*. The

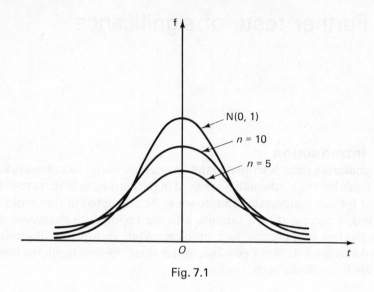

Fig. 7.1

degrees of freedom, v, of a statistic is the number of independent readings in the sample, that is n minus the number of population parameters that have had to be estimated from the sample. Hence, in general, $v = n - 1$.

Example 1 A training programme for word-processor operators claims to increase their speed by 10 words per minute. A random sample of 16 operators undertake the training and show an average increase of 6 words per minute with estimated population variance $\hat{\sigma}^2 = 81$ (words per minute)^2. Assuming that the increase in words per minute is normally distributed, investigate whether or not the sample results support the claim.

We cannot use z and the normal area tables here because n is small and σ is not known.

$H_0 : \mu = 10$,
$H_1 : \mu < 10$, one-tailed, since we are interested only in the *increase* being less than 10 words per minute.

We will use a 5% level of significance.

$$t = \frac{(6 - 10)}{9/\sqrt{16}} = -1\cdot778.$$

The degrees of freedom $v = 16 - 1 = 15$.

The t table for $v = 15$ gives the critical value for t, one-tailed, 5%, equal to $-1\cdot75$. That is, $P(t < -1\cdot75) = 0\cdot05$.

Since $-1\cdot778 < -1\cdot75$ (see Fig. 7.2), we must reject H_0 at the 5% level of significance. The sample results do not support the claim.

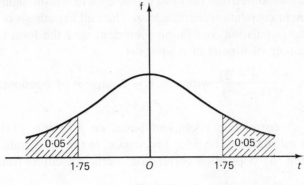

Fig. 7.2

Example 2 In a production process, bags of sugar are supposed to contain an average of 1 kg of sugar. A random sample of 9 bags is taken from the production and is found to have a mean content of 1·03 kg with population variance estimate of $(0·045)^2$ $(kg)^2$. Assuming that the weights of the contents of the packets are normally distributed, investigate whether or not this indicates that, at the 5% significance level, the mean content of the bags being produced is significantly different from 1 kg.

$H_0 : \mu = 1$,
$H_1 : \mu \neq 1$, two-tailed, since we are concerned with a change in either direction.

$$t = \frac{(1·03 - 1)\sqrt{9}}{0·045} = 2, \qquad v = 9 - 1 = 8.$$

Tables give the 5% significance level for t, two-tailed, when $v = 8$ as $t = \pm 2·31$ (see Fig. 7.3). Since $-2·31 < 2 < 2·31$, it follows that $t = 2$ is not within the region of rejection and we cannot reject H_0. The sample results do not indicate that the mean content is significantly different from 1 kg.

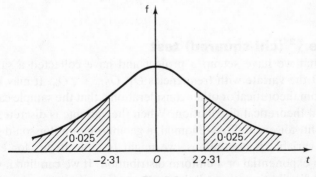

Fig. 7.3

In Chapter 6 we stated that the t-test can be used to test the significance of a product moment correlation coefficient, r. The null hypothesis is $H_0 : \rho = 0$, where ρ is the population correlation coefficient, and the form of the t-test used for a sample of n pairs of readings is

$$t = \frac{r \sqrt{(n-2)}}{\sqrt{(1 - r^2)}}, \text{ with } \upsilon = n - 2 \text{ degrees of freedom.}$$

Example 3 A correlation coefficient based on a sample of 16 pairs of readings was calculated to be 0·52. Investigate, at the 5% significance level, whether or not the population correlation coefficient differs from zero.

$$H_0 : \rho = 0, \qquad H_1 : \rho > 0.$$

$$t = \frac{0 \cdot 52 \times \sqrt{(16 - 2)}}{\sqrt{(1 - (0 \cdot 52)^2)}} = 2 \cdot 28, \qquad \upsilon = 14 \text{ degrees of freedom.}$$

Critical value of t at the 5% level for $\upsilon = 14$, one-tailed, is 1·7613. But 2·28 > 1·7613 and hence we reject H_0 at the 5% level. Yes, the population correlation coefficient does differ from zero.

Exercise 7.2

1 A correlation coefficient is calculated for a sample of 18 pairs of readings. Its value is found to be 0·46. Investigate whether or not this leads you to conclude (a) at the 5%, (b) at the 1% significance level, that there is positive correlation between the two variables.

2 Test whether the following random sample of 10 observations could reasonably have been taken from a normal distribution whose mean was 0·5.

1·1 2·3 1·2 0·0 0·9 1·7 0·7 4·4 1·3 1·2

3 Eight random samples are taken of a certain make of custard powder giving percentages of a certain ingredient as shown below. Find a 95% confidence interval for the mean percentage of that ingredient in that make of custard powder, assuming that percentages of that ingredient are normally distributed.

6·5 7·2 2·9 4·5 4·8 6·4 3·4 5·2

7.3 The χ^2 (chi-squared) test

Suppose that we have set up a project and have collected a sample of n readings of the variate with frequencies $O_1, O_2, \ldots, O_n$. It may be that we believe, from theoretical or other considerations, that the sample can be fitted by a known theoretical distribution. When the variable is discrete, we might be able to fit, for example, a binomial, a geometric or a Poisson distribution; when the variable is continuous, we might be able to fit, for example, a normal, an exponential or a uniform distribution. If we can find a well fitting theoretical distribution, then all the results we found for that distribution can be applied to the project. However, we cannot expect the collected data to fit exactly the theoretical distribution. We have collected *observed frequencies,* O_r, $r = 1, 2, \ldots, n$, that is, n *cells,* and we can calculate the *expected frequencies,* E_r, $r = 1, 2, \ldots, n$, which would occur under the particular theoretical distribution which we believe fits the collected data. The *statistic* χ^2 measures the discrepancy existing between the observed and the expected frequencies. We have to find the level of this measure at which we say this discrepancy is too large and therefore at which we cannot accept that the collected results are fitted by the chosen theoretical distribution. The discrepancy measure, χ^2, is given by

$$\chi^2 = \frac{(O_1 - E_1)^2}{E_1} + \frac{(O_2 - E_2)^2}{E_2} + \ldots + \frac{(O_n - E_n)^2}{E_n}$$

$$= \sum_{r=1}^{n} \frac{(O_r - E_r)^2}{E_r}.$$

The equation of the distribution of χ^2 is known; the shape of the distribution curve is dependent on the number of cells, and the curve is not symmetrical. Tables of percentage points for the χ^2 distribution are published showing the critical values of χ^2 for various significance levels and for values of v, the degrees of freedom. For a sample of n observed frequencies, $O_1, O_2, \ldots, O_n$, of n values of the characteristic being recorded (that is, there are n cells), where no population parameters have had to be estimated from the observed data in order to find the expected frequencies, then $v = n - 1$. When k population parameters have had to be estimated from the recorded data in order to find the expected frequencies, then $v = n - 1 - k$.

We stated that we assume that the theoretical distribution does fit the recorded data and that then we find the discrepancy between the observed and the expected frequencies. This assumption is H_0 in the significance test.

Example 4 Genetic theory states that certain cross breeding produces offspring of types A, B, C and D in the ratios $1:3:3:1$ respectively. In an experiment, 200 crosses gave the numbers of offspring of the four types as 18, 64, 82 and 36 respectively. Investigate whether these experimental results differ significantly from the theory.

H_0: the distribution of offspring of types A, B, C and D is in the ratios $1:3:3:1$, that is, the theory does fit the observed results.
Under H_0, out of 200 crosses 1/8 should be of type A, 3/8 of type B, and 3/8 of type C. We can then find the number of type D by subtraction from 200.

	A	B	C	D	Total
O	18	64	82	36	200
E	25	75	75	25	200

We have 4 cells so there will be 4 terms involved in calculating χ^2, and

$\upsilon = 4 - 1 = 3$. (You will notice that we had to calculate 3 of the expected frequencies.)

$$\chi^2 = \frac{(18 - 25)^2}{25} + \frac{(64 - 75)^2}{75} + \frac{(82 - 75)^2}{75} + \frac{(36 - 25)^2}{25}$$

$$= \frac{49}{25} + \frac{121}{75} + \frac{49}{75} + \frac{121}{25}$$

$$= 9 \cdot 07.$$

From χ^2 tables we have that the critical value of $\chi^2_{0 \cdot 05}$ (that is, the 5% level), when $\upsilon = 3$, is $7 \cdot 81$. Since $9 \cdot 07 > 7 \cdot 81$, P < 5% (see Fig. 7.4), and we reject H_0. The observed results differ significantly, at the 5% level, from the theory.

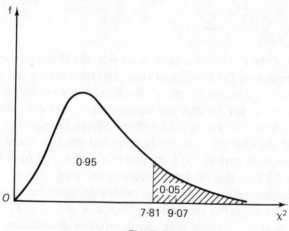

Fig. 7.4

From the formula for χ^2 it is obvious that χ^2 will be zero if there is perfect agreement between the two sets of frequencies, observed and expected. The larger the differences between the two sets become, the larger will be the value of χ^2. However, if an expected frequency is very small when compared with the other expected frequencies, the term $(O - E)^2/E$ which it contributes to the value of χ^2 will be very large; this is out of all proportion to the importance of that cell, since it has a very small expected frequency. The usual practice is to take any cell for which the expected frequency is less than 5 and to pool that cell with an adjacent cell.

Example 5 The numbers of pages with k misprints in a book of 500 pages were counted and the results are shown below. Calculate the expected frequencies under a Poisson distribution. Use a χ^2 test to decide whether or not the observed results differ significantly from those expected under the Poisson distribution.

Number of misprints (k)	0	1	2	3	4	Total
Observed number of pages	250	170	60	15	5	500

Since the probability of a misprint is small and the number of pages is large, a Poisson distribution is a suitable distribution to try to fit to the data. In order to find the expected frequencies, we need to estimate the mean μ of the distribution. From the observed results we have

$$\bar{x} = \hat{\mu} = \frac{0 \times 250 + 1 \times 170 + 120 + 45 + 20}{500} = 0 \cdot 71.$$

We can evaluate the first four expected frequencies using

$P(X = r) = \dfrac{e^{-\mu}\mu^r}{r}$, for $r = 0, 1, 2, 3$, using $0 \cdot 71$ as an estimate of μ.

$P(X = 0) = e^{-0 \cdot 71} = 0 \cdot 4916 \Rightarrow E_0 = 500 \times 0 \cdot 4916 = 245 \cdot 8.$

$P(X = 1) = 0 \cdot 71 \times P(X = 0) \Rightarrow E_1 = 0 \cdot 71 \times 245 \cdot 8 = 174 \cdot 5.$

$P(X = 2) = \dfrac{0 \cdot 71}{2} \times P(X = 1) \Rightarrow E_2 = \dfrac{0 \cdot 71}{2} \times 174 \cdot 5 = 62 \cdot 0.$

$P(X = 3) = \dfrac{0 \cdot 71}{3} \times P(X = 2) \Rightarrow E_3 = \dfrac{0 \cdot 71}{3} \times 62 \cdot 0 = 14 \cdot 7.$

$$E_4 = 500 - E_0 - E_1 - E_2 - E_3 = 3.$$

k	0	1	2	3	4	Total
O_r	250	170	60	15	5	500
E_r	245·8	174·5	62·0	14·7	3·0	500

The value E_4 is less than 5 so we combine the last two cells to give

k	0	1	2	$\geqslant 3$	Total
O_r	250	170	60	20	500
E_r	245·8	174·5	62·0	17·7	500

The degrees of freedom $v = 4 - 1 - 1 = 2$, since we had to find the estimate of μ from the collected data.

H_0: the Poisson distribution does fit the observed data.

$$\chi^2 = \frac{(4 \cdot 2)^2}{245 \cdot 8} + \frac{(4 \cdot 5)^2}{174 \cdot 5} + \frac{(2)^2}{62} + \frac{(2 \cdot 3)^2}{17 \cdot 7} = 0 \cdot 551.$$

This value of χ^2 is very small, but, nevertheless we refer to the tables to confirm our conclusion. The 5% critical value of χ^2 is 5·99. Because 0·551 is much less than 5·99, we cannot reject H_0. The observed results do not differ significantly from those expected under a Poisson distribution.

A useful application of the chi-squared test is in the testing of *contingency tables*. In this type of problem, the sample and the population are classified for more than one attribute. We restrict our work to two attributes, but the method is very similar when more attributes are recorded. We illustrate the method by an example of a 2×2 *contingency table*.

Example 6 Hoya cuttings are taken and grown in growing bags. Of 100 cuttings, 60 were placed straight into the bag, 40 were first dipped in a rooting powder. After 4 weeks of identical growing conditions, the cuttings were removed from the bags and the amount of root that each cutting had produced was classified as 'good' or 'poor'. The results are shown below:

Rooting

Treatment	Good	Poor
None	29	31
Rooting powder	26	14

Investigate whether these results lead you to conclude that the rooting powder affected the ability to root of the cuttings.

H_0: the rooting powder does not affect the ability to root. Under H_0, that is, 'no treatment' or 'rooting powder' have the same effect on rooting, 55 out of 100 have 'good' roots. Thus, out of the 60 which had no treatment, we would expect $\dfrac{55}{100} \times 60 = 33$ to have 'good' roots. All the other expected frequencies can now be found by subtraction from the totals. We write each expected frequency in brackets beside the corresponding observed frequency:

Rooting

Treatment	Good	Poor	Total
None	29(33)	31(27)	60
Rooting powder	26(22)	14(18)	40
Total	55	45	100

Since only *one* expected value needed to be calculated (and all the others followed by subtraction from the totals) we have $v = 1$.

$$\chi^2 = \frac{4^2}{33} + \frac{4^2}{22} + \frac{4^2}{27} + \frac{4^2}{18} = 2 \cdot 69.$$

The 5% critical value of χ^2 is $3 \cdot 84$. Since $2 \cdot 69 < 3 \cdot 84$, we cannot reject H_0. We conclude that the rooting powder does not significantly affect the rooting ability.

Note: Although in this example the expected frequencies were all integers, this will not be so in most examples. Since these are only *expected* frequencies, there is no need for them to be whole numbers and we keep one or two decimals according to the accuracy we require.

Exercise 7.3

1 Children having one parent of blood type A and the other of blood type B are expected to be of types A, AB and B in the ratios $1:2:1$. Of 200 such children, 42 are found to be of type A, 64 of type B and the remainder of type AB. Use the χ^2 test with a 5% significance level to discuss the agreement of these results with the theory.

2 In an outbreak of a disease in cattle, of 120 animals which were not inoculated, 20 caught the disease. Of 90 inoculated animals, 5 caught the disease. Use a χ^2 test to investigate the effectiveness of the inoculation shown by these results.

3 180 patients who complained of backache were divided into two groups, A and B. Group A, 58 people, were given a drug and, as a result, 46 said that their backache improved, 12 said that it did not. Group B, 122 people, thought that they were given the drug but, in fact, they were given a pill which contained no drug (a placebo). Of these, 88 said their backache improved, 34 said it did not. Perform a χ^2 test to investigate whether the drug was more effective than a placebo.

Miscellaneous Exercise 7

1 Minor earthquakes within a given intensity range occur each day in the vicinity of an ocean trench with the frequencies shown below:

Number of earthquakes per day	0	1	2	3	4	5	6
Frequency in days	11	20	23	13	4	1	0

Find the mean number of earthquakes per day.

Find, to one decimal place, the corresponding expected frequencies under a Poisson distribution with the same mean. Test the two frequency distributions for goodness of fit using a χ^2 test with 5 cells and 3 degrees of freedom.

2 A sample of 10 observations drawn at random from a normal population has mean 1·54 and population variance estimate 1·1513. Find 95% confidence limits for the true value μ of the population mean.

3 Corresponding values of the two variables x and y in a trial were

x	4	4	5	5	6	6	7	7
y	6	9	8	10	8	9	10	8

Find the equation of the line of regression of y on x.

Find the correlation coefficient, and, using a t-test, discuss its significance.

4 In a survey of 1000 families with 2 children, it was found that 270 families had 2 girls, 492 had a boy and a girl, and 238 had 2 boys. Investigate, at the 5% level, whether these figures are consistent with the hypothesis that boys and girls in families of two children are distributed binomially with $p = \frac{1}{2}$.

5 According to theory, the offspring of a certain cross in mice are in the ratios $7:3:5$ for black, brown and white offspring respectively. In an experiment it is found that, of 60 offspring produced as a result of the cross, 30 are black, 8 are brown and 22 are white. Use a χ^2 test to decide whether or not these sample results support the theory.

6 A chocolate bar manufacturer asked 120 children to choose one bar out of 6 different types of small chocolate bar which he manufactured. The results are shown below. Use a χ^2 test to check the hypothesis that there is no preference between the 6 types of chocolate bar.

Bar type	1	2	3	4	5	6	Total
Frequency	12	25	30	13	27	13	120

7 The sexes of puppies in 128 litters of 5 offspring per litter were recorded and the number of males is shown below:

Number of males in litter	0	1	2	3	4	5
Number of litters	8	22	30	40	18	10

Taking the null hypothesis that there is a 1:1 sex ratio in puppies, set up a theoretical distribution.

Use a χ^2 test to determine whether, on the basis of these data, the null hypothesis should be rejected.

8 Random samples are taken of 300 adults living in a large city and 200 adults living in the country. The numbers from each sample who smoke are shown below. Investigate, stating clearly the significance level you are taking, whether these results show a significant difference in the numbers who smoke for people living in the city and in the country.

	Smoke	Do not smoke
City	114	186
Country	82	118

9 Production records for a machine producing steel pins show that the pins have a mean length of 2·3 cm. On a certain day, to test the working of the machine, a random sample of 10 pins from the production is taken. This sample has mean length 2·26 cm with standard deviation 0·04 cm. Investigate whether or not this leads you to conclude, (a) at the 5%, (b) at the 1% significance level that the machine is working properly that day.

10 In a test of driving skills, 6 racing drivers take a reaction test for emergency braking. Their results, in seconds, are as follows:

$$0·31 \quad 0·29 \quad 0·46 \quad 0·57 \quad 0·42 \quad 0·34.$$

Find (a) 95%, (b) 99% confidence limits for the mean reaction time of racing drivers, assuming that reaction times are normally distributed.

11 The specification for the content of artificial colouring in a food product is that there is a maximum content of 1·1%. A sample of 12 analyses of this product shows that the sample has a mean artificial colouring content of 1·13% with a standard deviation of 0·05%. Given that production of the food must stop if the mean artificial colouring content is above 1·1%, use a t-test to decide whether or not, at the 5% significance level, this sample leads you to conclude that production must stop.

8 Project work

Project work is a compulsory part of the syllabus of some A-level boards but not for others. However, since statistics is essentially an experimental subject, practical or project work is an integral part of the teaching of statistics at A-level whether or not practical work is required for the actual examination. Indeed, the ability to apply the subject is essential to a statistician. The study of the theory, and the knowledge of statistical formulae is not enough; students must have some experience of the practical difficulties of collecting data, analysing it correctly by applying the correct methods, and interpreting the results in the light of the method and the problems of the data collection. Sometimes the data appear to contain 'errors' but, nevertheless, students must try to deduce sensible conclusions from collected data. They will find that, unlike most of the data given in class exercises, data collected in practice are not always 'ideal'. Projects will, hopefully, develop in students the ability to follow through the steps below in a logical way.

1 Problem identification We cannot start until the problem that we are trying to answer is properly defined, and the variable or statistic involved in the problem is clearly understood.

2 Design of the project Here several questions must be answered. We must decide whether the question involves interval estimation or hypothesis testing; also, how many random samples we require and whether small or large samples are suitable. We must also decide what assumptions are required for the theory we intend to use, and whether those assumptions are justified.

3 Collection of data Ideally this should be of data collected from students' own interests or from experiments performed in other subjects, for example, biology, geography and geology. This exposes them to the difficulties encountered in data collection and also helps them to establish relationships between collected data and theoretical distributions.

Data may be collected by observation or measurement, by interview, by postal questionnaire or by taking it from published statistics. Observed and measured data are probably the most satisfactory, provided that general care is taken, and measuring instruments are checked for accuracy.

Interviews and postal questionnaires both consist of a set of questions. These should be simple, unambiguous questions forming a short, concise questionnaire. There should be no leading questions (e.g. 'You agree, don't you, that ...?') and no irrelevant questions. Even with an ideal set of questions, one cannot rely on true answers nor be sure that people understand the questions. In the case of a postal questionnaire, few will be returned and the sample that is returned may well be biased. For example, suppose a questionnaire is sent to girls who leave school at 18 in order to start work and this questionnaire asks them to state the type of job they have obtained and their salary 6 months after leaving school. Then experience shows that those girls who have found a job with a good salary are more likely to return the questionnaire than those who have had no success in finding a job or have a poorly paid one. Hence the sample of replies is biased. Similarly, if housewives owning a certain make of cleaner were asked questions about its reliability, those with a complaint to make are more likely to return the questionnaire than those who are quite satisfied with their cleaner and who may well not bother to return the form.

The ability to acquire published data of a suitable type requires a knowledge of how to write to appropriate firms, associations, or government departments. Experience indicates that industry and industrial associations are extremely helpful to students who write to them courteously and explaining clearly what statistics they hope to obtain. Published government figures are a useful source of data provided we have some understanding of the background to its collection.

4 Analysis of data The type of analysis which is suitable to the problem should be indicated by the theoretical class work that has been studied. Tables and diagrams must be properly labelled.

5 Conclusions and comments The presentation of the results must include the conclusions in words so that another reader, not necessarily a statistician, can understand the purpose and results of the project. For example, '$z = 2.83**$' is not good enough; this would not be intelligible to most readers.

Class discussion of completed projects, led if necessary by a teacher, can be valuable. In this situation the writers can develop the ability to explain and even to defend their completed work. This can be a valuable asset when working in industry or taking part in a research seminar.

Some suggested topics for projects

1. The abilities of members of a group to estimate lengths or volumes or weights etc.

2. The distribution of weeds in a given area of lawn.

3. The effect of fertiliser, of temperature or of rainfall on plant growth.

4. Lengths of playing time of records in the pop charts.

5. Price variations, e.g. supermarket versus small shop, or price changes over a given period of time.

6. Age of motorcyclists and number of accidents.

7. Distribution of boys and girls in families of 2 or 3 children. Comparison with $p = \frac{1}{2}$ using large samples.

8. Distribution of errors when a given drawn line is trisected by eye. Relationship between absolute error and length of line.

9. Fit of a normal distribution to various body measurements.

10. Association between arm length and length around the wrist, or collar size.

11. Mean length of a large number of pencils and sampling distribution of means for large samples.

12. Sampling from Ordnance Survey maps, e.g. class A roads, churches, hotels.

13. Ages of coins in circulation.

14. Traffic passing a particular point on a road.

15. Times taken in getting to school; association between being late for school and poor transport service.

16. An investigation of football pool winnings and losses.

17. Distribution of birthday dates in a large gathering.

18. Ranking of advertisements, or photographs, or of sports stars.

19. Relationship between age and pulse-rate and prediction from this.

20. Income and expenditure amongst pupils.

21. Unemployment and gross national production.

22. Readership of newspapers and social classes.

23. Accidents per mile travelled in various years and in various forms of transport.

24. Death rates in various social classes.

25. Infant mortality rates in various parts of the country.

26. A study of truancy: by sex or by age. Connection between truancy and crime.

27. Connection between mock and final O level grades.

28. Comparison of a school's O and A level results with the corresponding results for the district.

29. A study of use of the school library; who uses it, what books are borrowed, length of time that books are borrowed, numbers of overdue books, etc.

30. A study of football results, or of rugby results, and of attendances at matches.

ANSWERS

Exercise 1.3
1 A suitable value would be to measure to 0·01 mm.
3 A suitable grouping would be 15 classes of width 0·11 mm.

Exercise 1.4
1 250

Exercise 1.5
1 ≈ 31·6 years
2 2·86
3 9 days, 0 days, 6 days
4 (a) 8·3 mg/litre, (b) 2·11 mg/litre, (c) 6·50 (mg/litre)^2, 2·55 mg/litre
5 169·85 cm; 166–
6 169·7 cm; 166·2 cm, 173·5 cm

Exercise 1.6
1 5·8 cm, 2·36 cm; 6 cm, 1·87 cm; 5·89 cm, 4·65 cm^2

Exercise 1.7
1 152; ≈ £1·29

Miscellaneous Exercise 1
1 16·85 flies, 6·89 flies; 17·5 flies, 22 flies
2 110·7, 685·51
3 ≈ 57 marks; ≈ 59 marks; ≈ 19 marks
4 ≈ 22·5 sec; ≈ 22·8 sec; ≈ 21·8 sec, 23·4 sec
5 13·21%, 1·31%
6 3·86, 1·12; 3·99, 0·94
7 131·4, 16·78, 13·15
8 (a) 104·24% (b) 104·33% (c) 104·29%
9 $x = 10, y = 10$
10 1·35, 0·872

Exercise 2.1
1 4, ≈ 2·28; 4, 1·95
2 5, 11·5; 5; 5·75

Exercise 2.2
2 0·023; 0·976
3 271
4 0·007

Miscellaneous Exercise 2
1 μ, σ^2/n; (a) approximately normal, (b) normal; 0·742, 6221 g to 6279 g
2 0·0401; ≈ 384 samples
3 126·2 samples
4 2·18%, 0·26%
5 0·6915; 0·0026

Exercise 3.1
1 1·36, 0·001
2 1·01 m, 0·008 m^2
3 620 hours, 22·14 hours

Exercise 3.2
1 0·2635
2 5·087 to 7·413
3 1·49 V, 0·0644 V; 0·0065 V; 1·477 V to 1·503 V
4 £28·05 to £28·75; 3119

Miscellaneous Exercise 3
1 0·0455
2 9·5; 2·6
3 246
4 28 kN, 18·29 (kN)^2; (28 ± 0·85) kN; 174
5 0·421; 2552
6 0·96
7 14·2, 0·041; (14·2 ± 0·08); 0·007
8 0·3108; 0·8664
9 (a) normal, (b) approximately normal; (411·63 ± 0·108) nm
10 (a) 0·113 (b) 0·763 (c) 3·79 g, (d) 0·137

Exercise 4.2
1 20·3 kg, 0·6633 kg; we cannot reject his claim.
2 (a) Reject the null hypothesis at

the 5% level. (b) We cannot reject the null hypothesis at the 1% level.

Exercise 4.3

1. We cannot reject the hypothesis that the customers like the cheeses equally well.
2. No; reject H_0 that the coin is fair.
3. (a) Yes; significant at the 2% level. (b) No; not significant at the 1% level.

Miscellaneous Exercise 4

1. (a) Yes; significantly large. (b) No; not significantly large.
2. Yes; significantly greater than the national average at the 5% level.
3. (a) 0·132, (b) 0·842; cannot reject the hypothesis that sex ratios are equal.
4. 84·75; 1·232 mm to 1·256 mm
5. $P(X \leq 3) = 0·225$; result supports the theory.
6. Cannot reject the null hypothesis at the 2% level.
7. (a) 0·029, (b) 0·412; yes, significant at the 1% level.
8. (a) 0·681, (b) 0·462; 4·47 cm to 5·29 cm; we cannot reject, at the 5% level, the hypothesis that the mean diameter is the same on the island and elsewhere.
9. 116; (a) 21·4%, (b) 20·7%; 0·292; no, results do not suggest that the sample is biased.
10. Significant increase at the 1% level.
11. 0·995; we conclude, at the 5% level, that the probability of a dose causing infection is less than 0·8.

Exercise 5.1

1. $y = 0·795x + 13·339; 82·504$
2. $R = 5·64C + 1·99; 16·09$ kg; 1·305

Miscellaneous Exercise 5

1. $y = 1·12x + 5·59$
2. $y = -0·088x + 3·471; 3·295$
3. $y = 0·17x + 3·20;$ 0·17 mg/1000 g/hour
4. $M = 0·329T + 41·2$
5. $D = 6·99 - 0·008A$
6. $e = 1·71W + 0·09; 1·71$ cm

7. $V = 17·35T - 114·83; -14·2$
8. $\bar{x} = 29·7, \bar{y} = 124; y = 2·65x + 45·21$
9. In $T = 0·0786p - 4·086;$ $T \approx 310$ g cm^{-2}
10. In $(10 - C) = 2·249 - 0·151t; m = 9·48, k = 0·151; C = 6·7$ mg l^{-1}

Exercise 6.1

1. 0·39
2. 17·56, 0·717
3. 0·813

Exercise 6.2

1. (a) $r_S = 0·5$, (b) $r_K = 0·\dot{3}$
2. $r_S = 5/6, r_K = 9/14$
3. $r_S = 0·676, r_K = 0·5\dot{3}$

Miscellaneous Exercise 6

1. $0·689; r_S = 0·703$
2. 0·982
3. 0·663; 0·657
4. $r_S = 0·943, r_K = 0·867$
5. $Y = 1·323X - 2·161, X = 0·478Y + 6·915, r = 0·795$
6. $0·720; r_S = 0·810, r_K = 0·643$
7. $r_S = 0·7, r_K = 0·6$
8. 0·688
9. $r_S = 0·912, r_K = 0·756$
10. $-0·741; y = 16·265 - 0·649x; \approx 7$ mistakes

Exercise 7.2

1. $t = 2·072;$ (a) significant, (b) not significant.
2. $t = 2·61;$ no, reject H_0 at the 5% level.
3. 3·839% to 6·386%

Exercise 7.3

1. $\chi^2 = 5·56;$ we consider results fit the theory.
2. $\chi^2 = 6·04;$ inoculation does appear to be effective.
3. $\chi^2 = 1·05;$ drug and placebo appear to be equally effective.

Miscellaneous Exercise 7

1. 1·75; 12·5, 21·9, 19·2, 11·2, 4·9, 1·7, 0·6; $\chi^2 = 2·06$
2. 0·77 to 2·31
3. $y = 0·4x + 6·3; 0·365; t = 0·96$
4. $\chi^2 = 2·304;$ yes, results are consistent.
5. $\chi^2 = 1·68;$ yes, results support the theory.

6 $\chi^2 = 16 \cdot 8$; reject H_0.

7 4, 20, 40, 40, 20, 4; $\chi^2 = 4 \cdot 6$; we cannot reject H_0.

8 $\chi^2 = 0 \cdot 453$; no apparent difference shown.

9 $t = 3$; (a) reject H_0, (b) accept H_0.

10 $0 \cdot 287$ s to $0 \cdot 510$ s; $0 \cdot 223$ s to $0 \cdot 574$ s.

11 $t = 1 \cdot 99$; conclude that production must stop.

Index